高等院校土建学科双语教材（中英文对照）
◆ 建筑学专业 ◆
BASICS

工程制图
TECHNICAL DRAWING

[德] 贝尔特·比勒费尔德
[西] 伊莎贝拉·斯奇巴   编著
吴寒亮　何玮珂　译

中国建筑工业出版社

著作权合同登记图字：01-2007-3336号

图书在版编目（CIP）数据

工程制图/（德）比勒费尔德，（西）斯奇巴编著；吴寒亮、何玮珂译.
北京：中国建筑工业出版社，2001
高等院校土建学科双语教材（中英文对照）◆建筑学专业◆
ISBN 978-7-112-11582-2

Ⅰ.工… Ⅱ.①比…②斯…③吴…④何… Ⅲ.工程制图-高等学校-教材-英、汉 Ⅳ.TB23

中国版本图书馆 CIP 数据核字（2009）第 209814 号

Basics：Technical Drawing/Bert Bielefeld，Isabella Skiba
Copyright © 2007 Birkhäuser Verlag AG（Verlag für Architektur），P. O. Box 133，
4010 Basel，Switzerland
Chinese Translation Copyright © 2011 China Architecture & Building Press
All rights reserved.
本书经 Birkhäuser Verlag AG 出版社授权我社翻译出版

责任编辑：孙　炼
责任设计：郑秋菊
责任校对：袁艳玲　赵　颖

高等院校土建学科双语教材（中英文对照）
◆ 建筑学专业 ◆
工程制图
[德] 贝尔特·比勒费尔德
[西] 伊莎贝拉·斯奇巴　编著
吴寒亮　何玮珂　译
*
中国建筑工业出版社出版、发行（北京西郊百万庄）
各地新华书店、建筑书店经销
北京嘉泰利德公司制版
北京密东印刷有限公司印刷
*
开本：880×1230 毫米　1/32　印张：4¼　字数：136 千字
2011 年 5 月第一版　2011 年 5 月第一次印刷
定价：**16.00** 元
ISBN 978-7-112-11582-2
(20282)

版权所有　翻印必究
如有印装质量问题，可寄本社退换
（邮政编码 100037）

中文部分目录

\\ 序 5

\\ 投影分类 77
 \\ 俯视图（屋顶平面图） 77
 \\ 平面图 77
 \\ 立面图 78
 \\ 剖面图 79
 \\ 三维视图 79

\\ 表现原则 81
 \\ 辅助工具 81
 \\ 图纸规格和类型 82
 \\ 比例 85
 \\ 图形填充 88
 \\ 文字标注 89
 \\ 尺寸标注 90

\\ 制图步骤 95
 \\ 定项基础 95
 \\ 初步设计图 97
 \\ 表现图 99
 \\ 设计图纸 105
 \\ 设计许可 111
 \\ 施工图 115
 \\ 专业图纸 124

\\ 图纸的表现方法 126
 \\ 图纸的组成 126
 \\ 图签 126
 \\ 图纸布局 127

\\ 附录 130
 \\ 符号 130
 \\ 标准 133

CONTENTS

\\Foreword _7

\\Projection types _9
    \\Top view (or roof plan) _9
    \\Plan view _9
    \\Elevation _10
    \\Section _10
    \\Three-dimensional views _11

\\Principles of representation _14
    \\Aids _14
    \\Paper formats and paper types _16
    \\Scale _18
    \\Lines _19
    \\Hatching _21
    \\Labelling _22
    \\Dimensioning _24

\\Planning stages _30
    \\Determining basics _30
    \\Preliminary design drawing _30
    \\Presentation plans _39
    \\Design planning _42
    \\Planning permission _50
    \\Working plans _52
    \\Specialist planning _63

\\Plan presentation _66
    Plan composition _66
    Plan header _66
    Plan distribution _68

\\Appendix _71
    \\Symbols _71
    \\Standards _74

序

设计是建筑物建造的必需条件。首先，建筑物必须在方案图和施工图上呈现出来。通常情况下，设计的第一阶段是方案草图设计，如设计简图或者透视图，其目的是确定建筑物的形式和设计方案。本书介绍的内容是在第一阶段已完成之后才开始进行的工作。在接下来的时间里，首先是将设计简图转化成具有精确几何比例的设计图：工程图纸。工程图纸描述了在设计或者施工过程中将出现的情况及其详细过程，是建筑物建造过程中必不可少的资源。

这套基础教材旨在从教学目的出发，以一种适应练习的方式来给出内容。我们将向读者介绍建筑学中不同专业领域的培训内容，并以一种简洁、系统的方式来讲述这些基础知识。

本书面向建筑学和土木工程专业的低年级学生，以及初学建筑制图和工程制图的读者，尤其是大学课程对工程制图要求进行相关基础知识的准备，而学生必须通过自学来得到这些基础知识。其中的难点在于规范建筑制图的大量 ISO 标准，而这些国际标准大部分是基于德国的 DIN 标准（例如纸张的版式）。然而，如果不注意这些通常的规则，就没有正确的途径来准备和开展设计或者制图工作。建筑制图始终是个人设计的一种自我表现，具有个人的格调。

因此，本书给出了在设计过程中不同的设计类型和制图方式的基本要求。掌握了这些要求，学生们就能够通过建筑制图快速、自信地展示他们的设计和思路。

编者：贝尔特·比勒费尔德

FOREWORD

Buildings are not erected without plans, but must first be presented in plans and construction drawings. As a rule, the first steps consist of free presentations like sketches or perspective drawings, intended to establish the form and design of the building. This book begins at the point when these ideas have been developed, and the first sketches are turned into geometrically precise scale plans: technical drawings. These drawings provide an image of what will emerge in terms of design or construction, and their detailing process is thus an essential resource on the way to a finished building.

The "Basics" series of books aims to present information didactically and in a form appropriate to practice. It will introduce students to the various specialist fields of training in architecture, and transmit the basics in a compact and systematic way.

The "Technical Drawing" volume is directed at those commencing studies in architecture and civil engineering, and trainees in construction drawing and technical drawing. University courses in particular often require basic knowledge for the preparation of technical drawings, which students have to acquire laboriously for themselves. The difficulty lies in the large number of ISO standards that regulate construction drawings – the German DIN standards were largely the basis for international standards (e.g. for paper formats). Regardless of these general rules, however, there is no one correct way of preparing and creating a design or working drawing. Construction drawings always an act of self-expression by the person preparing them; they have a personal touch.

This book therefore provides the fundamentals required for the various plan types and drawings in the planning process – so that students are enabled to present their designs and ideas quickly and confidently through construction drawings.

Bert Bielefeld
Editor

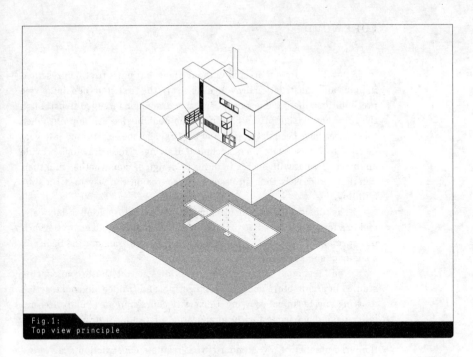

Fig.1:
Top view principle

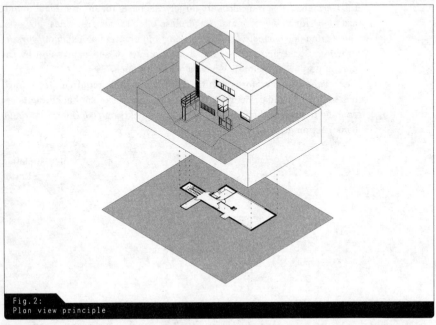

Fig.2:
Plan view principle

PROJECTION TYPES

Buildings are drawn in the form of sketches and free perspectives, but also using various structural drawing approaches. A fundamental distinction is made between top view and elevation for the exterior, and between plan view and section when drawing the interior of a building.

TOP VIEW (OR ROOF PLAN)
Top view drawings present a view or projection of the building seen from above. A top view (also often called roof plan) is important for the location plan, for example, which defines the building's position on the plot.

PLAN VIEW
In the same way, a plan shows a single floor of the building. Here a section is taken through the building at a height of about 1 to 1.5 m above the floor, to include as many apertures (doors, windows) in the masonry as possible. To make the drawing comprehensive, the heights of all the relevant structural sections (sill to floor height, aperture height, ground level, floor height) are given, as well as all the relevant horizontal dimensions. The height of the horizontal section may be changed to illustrate as many special features of the design as possible and to represent any windows that may, for example, be higher. (The different position of the window is then clarified by giving the sill to floor height.) For the direction of the plan, there are two basic possibilities:

_ The downward direction of view used in architects' plans (top view), which makes it possible to record room structures, form and size.
_ The reflected plan views from below showing construction elements that lie above the horizontal section level. – Structural engineers prefer this view, which shows the loadbearing construction elements in the ceiling above. ⟩ see chapter Specialist planning

Designating plan views
⟩ 📖

Plan views are generally designated according to the floor they apply to, e.g. cellar floor plan view, ground floor plan view, 1st floor plan view, attic floor plan view etc. If it is not possible to identify floors clearly in a design, e.g. when floor levels are offset, the obvious thing to do is to name the plan views after particular levels: e.g. level -3 plan view, underground car park.

> \\Hint:
> As a rule, plans with top and plan views are "northed", in other words, north is at the top edge of the plan and is indicated by a north arrow.

ELEVATION

Elevations (also called views in the ISO standards) show the outside of the building with all its apertures. Views of the cubature of a building provide information about its relationship to its environment, its form and proportions, and the construction type and material qualities where applicable. Along with the plan views and sections, elevations complete the overall design.

Elevations are parallel projections, seen from the side, onto a building façade. The projection lines run at right angles to the projection plane, so offset sections are not shown in their true size.

An elevation generally shows the immediate surroundings, with the lie of the terrain and links to any existing building development where appropriate.

Designating elevations

Elevations are identified according to their position on a point of the compass. The north arrow on the location plan and on the plan views defines the orientation of the building. Hence, the following designations are used for each of the four elevations: view north, south, east and west (or northeast, southwest etc.). If only two elevations are visible (as in terraced houses), they can also be defined in relation to the building's position on the plot, or the position of the development as a whole. But this means that only two sides are fixed unambiguously, e.g. the garden or courtyard side and the street side. The labelling of the elevations must be clear for anyone – even if they are not familiar with the area.

SECTION

A section is created by making a vertical cut through a building and considering this as a view in parallel projection. Sections are intended to provide information about floor heights, material quality and the building materials to be used for the planned building.

The section line must be entered on the plan view or all plan views. It is identified by a thick dash-dot line and the direction of view. Arrows and two capital letters of equal size fix the direction and the designation of the section. The section is taken in such a way that all the information

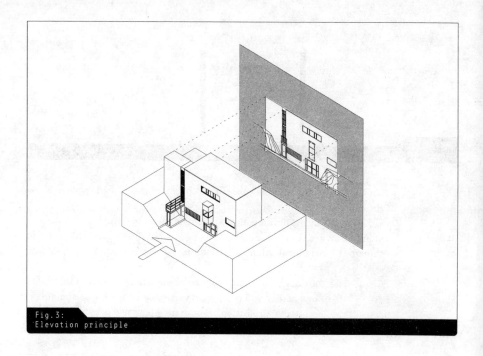

Fig.3:
Elevation principle

relevant to the building can be recorded, which means that the section line may deviate occasionally. This deviation must be at right angles, and must be identified in the plan view.

Elements of a section

Important elements shown in a section include the structure of the roof, the floors and ceilings, the foundations, and the walls with their apertures. The section should also show access to the building via stairs, lift, ramp etc.

Designating sections

Sections taken parallel through a main axis of the building are called longitudinal and cross sections. A longitudinal section cuts the building along the longer side and the cross section along the shorter. If more than two sections are taken, they are usually designated by capital letters or numbers. As the section lines on the plan views are identified by the same letters on both sides, the section designations are, correspondingly, section A–A, section B–B or section C–C etc.

THREE-DIMENSIONAL VIEWS

Axonometric projections

Axonometric projections are plan views or views with a third plane added – height. They are generally used as three-dimensional views at the planning stage, and give a spatial impression of the building. They are

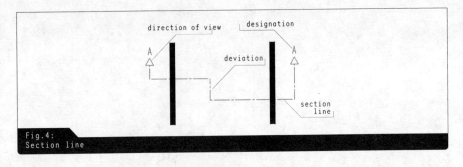

Fig.4:
Section line

used for construction plans only in exceptional cases, for example to show the design of a corner.

It is easy to develop three-dimensional views from a two-dimensional drawing. A distinction is made between different projections (even though the term "projection" is misleading here):

_ The "military projection", where the plan view is rotated through 45° at one corner and completed vertically by adding the heights.
_ The "architects' projection", where the plan view is again rotated at one corner – through 30° or 60°.

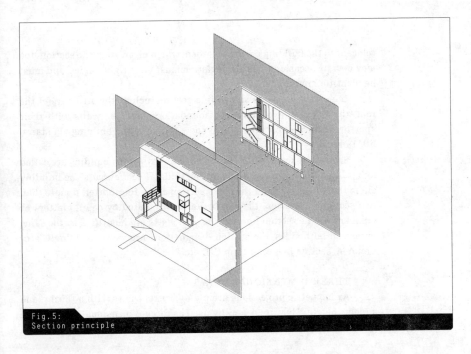

Fig.5:
Section principle

_ The "cavalier projection", where an elevation or section acquires a third dimension by the addition of lines at an angle of 45°.

Isometric and dimetric projections

It is possible to create a better three-dimensional impression by moving away from the right angle as a base. Isometric and dimetric projections are used to do this.

Isometric drawings place each of the two plan view axes at an angle of 30° to the horizontal base line, and the height axis is plotted onto the plan view axes. This means that the object represented is not so distorted as in the three methods given above, but the drawing must be constructed more elaborately.

For a dimetric projection the two plan view axes are placed at angles of 7° and 42°. The line lengths of the latter should be shortened by factors of 0.5 or 0.7.

Perspective drawings

Perspective drawings differ from axonometric, isometric and dimetric projections in that they do not present the lines lying on an axis as parallel, but in perspective. Since perspective drawings are not generally used as construction plans, but only for presentation purposes, they fall into the field of descriptive geometry and are not examined further here.

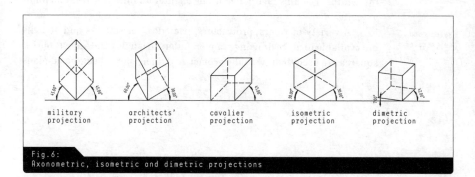

Fig.6:
Axonometric, isometric and dimetric projections

PRINCIPLES OF REPRESENTATION

AIDS

Basically there are two different methods for producing construction plans:

_ Hand drawing
_ CAD

Drawing tables

Hand drawing is carried out either at special drawing tables fitted with a pair of sliding rulers set at right angles, which can be adjusted; or using drawing rails, which are screwed onto an existing desk-top and slid vertically on stretched wires. Both variants make it possible to draw lines parallel or at right angles.

Pens and pencils for hand drawing

Hand drawings are usually made with pencils or ink pens. Pencils are available in various hardness grades, which affect the thickness and visual effect of a line: the harder the pencil, the finer the line, because little lead is rubbed off onto the paper. So various grades of pencil are needed for drawings, to be able to make lines of different widths.

>

Ink pens exist in various forms (e.g. with or without cartridges) and nib widths. The nibs mentioned in the chapter on lines are available individually.

>

Rulers and set squares

A variety of rulers, protractors, triangles, set squares and stencils are available to make drawing simpler. Rulers, triangles and adjustable set squares are used to draw in the geometrical dimensions. Lengths on plans

---

\\Tip:
Pencils ranging from grade B (soft) via F (medium) to H to 3H (various degrees of hardness) are used for construction plans. Harder pencils should be used first, to avoid smudging the softer, thicker lines.

\\Tip:
Ink pens come with and without cartridges in various colours. The latter are cheaper in throwaway versions, but more expensive if a lot of drawing is to be done. The thinner the line produced by an ink pen, the greater is the danger that the pen may dry up if stored for a long time. Sometimes the pigments in the nib of the pens can be moistened in a bath of water. If lines that have already been drawn need to be removed, special ink erasers can be used. But careful scratching with a razor blade is quicker.

Fig.7:
Typical aids

are generally measured with a set square, a triangular ruler including six different scales with a length scale for each.

Stencils

Stencils are available for almost all typical drawing symbols (e.g. for furniture, electrical connections or bathroom facilities). There are also stencils for standard typefaces. All stencils are dependent on line thickness and scale.

CAD programs

CAD drawings are made using a computer. You need a CAD (Computer Aided Design) program intended specially for construction drawing. Various programs are available on the market, but they differ considerably in ease of use, performance and price. Almost all providers offer student and school versions.

\\Important:
It makes sense to look at what several providers offer before working your way into a program. Prices vary considerably even between the student versions, and it is also important for fellow students or colleagues to use the same program, so that working experiences can be shared.

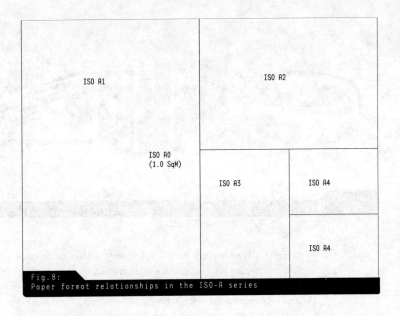

Fig. 8:
Paper format relationships in the ISO-A series

PAPER FORMATS AND PAPER TYPES

Paper formats

DIN 476-1 or ISO 216 define various paper formats based on a page ratio of $1:\sqrt{2}$. The advantage of this page ratio is that a large sheet can always be divided into smaller formats without waste.

There are various series within the DIN or ISO standards; the DIN-A or ISO-A series is generally used for plans.

As losses occur when cutting to size and folding, sheet formats distinguish between trimmed and untrimmed sheets. There are also format categories like DIN A3 Plus on the paper market, but these were created by printer manufacturers and are not further standardized.

Table 1:
ISO/DIN series A–E (mm × mm)

	A–	B–	C–	D–	E–
2-0	1189×1682	1414×2000			
0	841×1189	1000×1414	917×1297	771×1091	800×1120
-1	594×841	707×1000	648×917	545×771	560×800
-2	420×594	500×707	458×648	385×545	400×560
-3	297×420	353×500	324×458	272×385	280×400
-4	210×297	250×353	229×324	192×272	200×280

Table 2:
Untrimmed and trimmed DIN-A series papers (mm x mm)

DIN	untrimmed	trimmed	edge distance
2-A0	1230×1720	1189×1682	10
A0	880×1230	841×1189	10
A1	625×880	594×841	10
A2	450×625	420×594	10
A3	330×450	297×420	5
A4	240×330	210×297	5

The paper formats described above are recognized and used in most countries; but in North America in particular, some inch-based formats are used that are based on ANSI standards.

Table 3:
ANSI paper formats

Series	Engineers	Architects	Engineers	Architects
A	8½×11	9×12	216×279	229×305
B	11×17	12×18	279×432	305×457
C	17×22	18×24	432×559	457×610
D	22×34	24×36	559×864	610×914
E	34×44	36×48	864×1118	914×1219
F	44×68		1118×1727	
	in×in		mm×mm	

Paper types

Various paper types are distinguished, as well as paper formats. As a rule, tracing paper is used for hand drawings, as it has the advantage that other drawings can be placed underneath it to be traced. This considerably simplifies construction (e.g. of upper storeys or sections). Tracing paper also makes it possible to duplicate the original drawing simply, using blueprints.

In the inventory field, drawing films are often used, as they hold their shape even at higher temperatures, and thus measurements can be read off them reliably even after a long time.

For technical drawings made with CAD programs, normal white paper in roll or sheet form is usually used for plotting. Coated papers, or photographic or glossy papers are often used for presentation drawings, as they have a high-quality surface.

SCALE

Every type of plan mentioned in the first chapter (Projection types) is a reduction in a certain ratio to the built reality, i.e. it is drawn on a particular scale. The scale used must be marked on every drawing, in the form of the word scale and two figures separated by a colon (e.g. Scale 1:10).

Definition of scale

A scale describes the relationship between the dimensions of an element in a drawing and in the original size. A distinction is made between three principal scale types:

_ Original scale (scale 1:1) as the natural scale
_ Enlarged scale (scale x:1), in which one element is drawn larger than its natural size by a certain multiple
_ Reduced scale, (scale 1:x), in which one element is reproduced smaller than its actual size by a certain multiple

Thus, for example, a wall drawn on a scale of 1:100 will be one hundred times smaller than the original.

Typical scales

Reduced scales are almost always used for construction drawings, as the object represented is usually larger than the paper. As precision and detail in the design process increase, the reduction becomes less, so the object itself is represented as larger.

Location plans and rough surveys are often drawn on a scale of 1:500 (or 1:1000), design drawings on the scales of 1:200 or 1:100. For working plans, the scales of 1:50, 1:25, 1:20, 1:10, 1:5, 1:2 and 1:1 are used. › See chapter Planning stages

Converting scales

›✎

If a wall 5.5 m long is to be represented on the scale of 1:50, its length must be divided by the reduction factor: thus, 5.5 m/50 = 0.11 m. The length drawn is thus 11 cm. It becomes more difficult when an object that is already drawn on a reduced scale has to be converted to a different one. If a door with a drawn length of 5 cm at a scale of 1:20 is then shown on a scale of 1:50, the two scales must be calculated against each other. Thus, the length is 5 cm *20/50 = 5 cm/factor 2.5 = 2 cm.

Scales for CAD programs

CAD programs simplify the scale conversion problem. Here the building is usually input on a scale of 1:1, i.e. a wall 5.5 m long is drawn at this length. The drawing is additionally provided with an output or reference scale, which defines the scale on which the drawing will be printed and output later. Pen and lettering widths also adapt to this reference scale when viewed on the monitor, so that the ultimate result can be seen.

> **\\Tip:**
> To convert original dimensions into typical construction drawing scales it is best to use a scale set square, or calculate the length to scale as follows:
>
> _ Scales 1:10, 1:100, 1:1000 – move the decimal point by the number of noughts to change the scale unit from m to cm for 1:100 or mm for 1:1000
> _ Scales 1:5, 1:50, 1:500 – move the decimal point as above and then multiply the number by 2
> _ Scales 1:20, 1:200 – move the decimal point as above and then divide by 2

LINES

A technical drawing consists of lines that differentiate things according to their type and width. Here a distinction is made between line types and line widths, though their significance can vary from scale to scale.

Line types

The are four principal types of line: the unbroken line, the dashed line, the dash-dot line and the dotted line, and other intermediate forms can be developed from these.

Line widths

The following line widths are customary, although as a rule only widths up to 0.7 mm are used: 0.13 mm, 0.18 mm, 0.25 mm. 0.35 mm, 0.5 mm, 0.7 mm, 1 mm, 1.4 mm, 2 mm.

Using unbroken lines

The unbroken line is used for all visible objects and visible edges of building sections; boundaries of sectional areas are also identified by unbroken lines. When parts of a building are cut in sections on the scale of 1:200 and 1:100, unbroken lines 0.25–0.5 mm wide are generally used; on scales from 1:50 a width of 0.7–1 mm is recommended. Unbroken lines for auxiliary constructions, dimension lines or secondary top or plan views are drawn more finely: 0.18–0.25 mm wide for a scale of 1:200 or 1:100, and 0.25–0.5 mm from 1:50.

Using dashed lines and dotted lines

Dashed lines are used for concealed edges of building parts (e.g. the under-step in details of stairs) in line widths of 0.25–0.35 mm for scales of 1:200 and 1:100, and 0.5–0.7 mm for scales from 1:50.

Dash-dot lines define axes and section runs. As section runs need to be immediately recognized on the drawing, they are drawn at a line width of 0.5 mm for scales of 1:200 and 1:100, and 1 mm for scales from 1:50. Axes, on the other hand, are usually drawn in lines 0.18–0.25 mm wide for scales of 1:100 or 1:200, and 0.35–0.5 mm from 1:50.

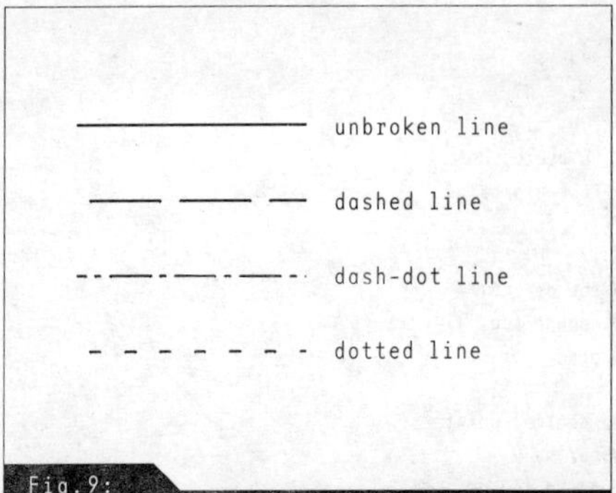

Fig.9:
Line types

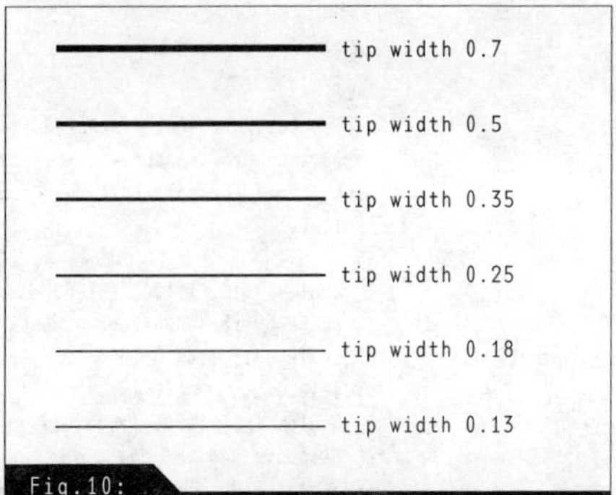

- ——————— unbroken line
- —— —— —— dashed line
- ·—·—·—·— dash-dot line
- — — — — — dotted line

Fig.10:
Line widths

- tip width 0.7
- tip width 0.5
- tip width 0.35
- tip width 0.25
- tip width 0.18
- tip width 0.13

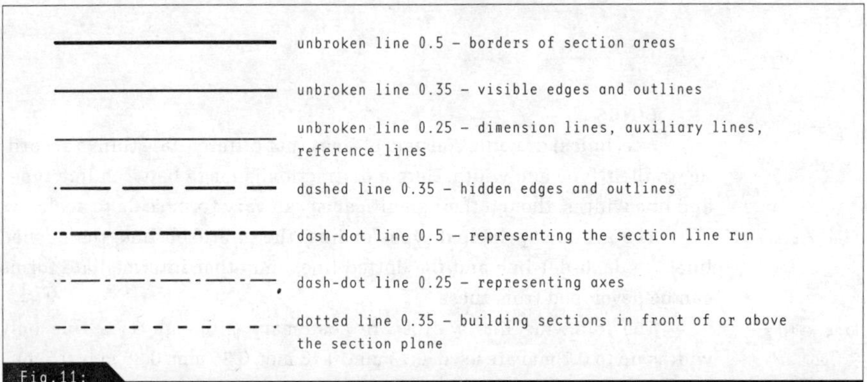

Fig.11:
Line types and widths, scale 1:100

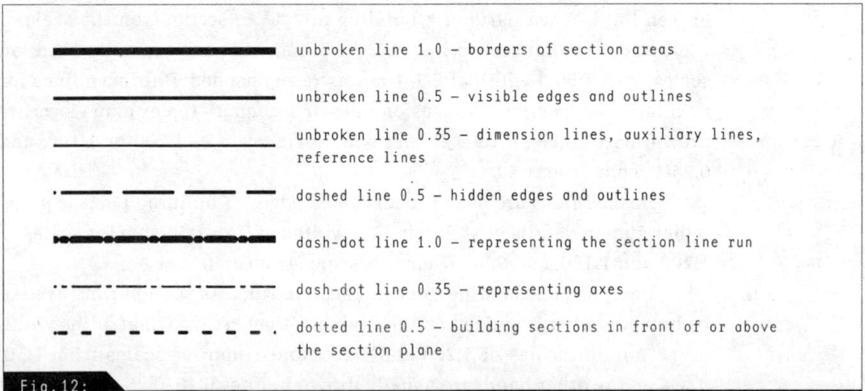

- unbroken line 0.5 – borders of section areas
- unbroken line 0.35 – visible edges and outlines
- unbroken line 0.25 – dimension lines, auxiliary lines, reference lines
- dashed line 0.35 – hidden edges and outlines
- dash-dot line 0.5 – representing the section line run
- dash-dot line 0.25 – representing axes
- dotted line 0.35 – building sections in front of or above the section plane

Fig.12:
Line types and widths, scale 1:50

- unbroken line 1.0 – borders of section areas
- unbroken line 0.5 – visible edges and outlines
- unbroken line 0.35 – dimension lines, auxiliary lines, reference lines
- dashed line 0.5 – hidden edges and outlines
- dash-dot line 1.0 – representing the section line run
- dash-dot line 0.35 – representing axes
- dotted line 0.5 – building sections in front of or above the section plane

Dotted lines identify the edges of building section that can no longer be represented because they are placed behind the section plane. › see also chapter Projection types Here a line width of 0.25–0.35 mm is used for scales of 1:100 and 1:200, and 0.5–0.7 mm from 1:50.

HATCHING

Hatching is intended to simplify representing individual elements in drawings, and to make them more intelligible. Hatching appears in section plans (plan views, sections) and provides information about the nature of the representation, and the qualities of the materials and components used in the planning. When sections are taken through parts of a building the lines around them are usually filled in with hatching. Most statements about the way hatching is presented have been summed up in national standards. › see Appendix Basically, there is a distinction between section areas that are not dependent on the materials, such as diagonal hatching or filler areas, and material-dependent representations. › see Fig. 13 A material-dependent representation identifies the material to be used for the part of the building through which the section has been taken. In the preliminary design phase, walls with a section through them are often shown only by filled areas of material-independent diagonal hatching on the plan view, to emphasize the solid parts of the building. Material-dependent hatching is not usually deployed until the working plan stage (e.g. masonry or reinforced concrete), as the appropriate materials will already have been chosen at this point.

Principles of hatching

Hatching can be presented as lines, dots, grids or geometrical figures. If the interfaces of several parts of the building are juxtaposed, then the direction of the hatching will change as well. Hatching is usually drawn at 45° or 135°.

\\Hint:
The above-mentioned line widths should be understood as guidelines, because of the current use of CAD. Today's CAD programs offer users pen categories that can be tailored to individual needs. So at times smaller line widths are used than in hand drawings. Test printouts should however be made at the beginning of a drawing to be able to estimate the effect of the line width on an output scale, as the on-screen effect often does not reflect the printed reality because of zoom functions and presentation that is independent of scale.

\\Hint:
In CAD programs hatching can be scaled in such a way that it makes visual sense for every output scale. In this context, the terms scale-dependent and scale-independent hatching are used.

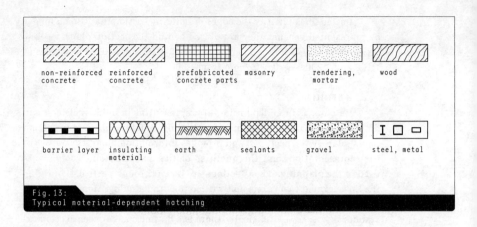

Fig. 13:
Typical material-dependent hatching

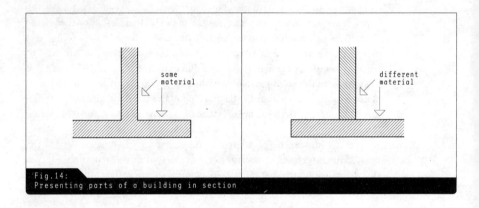

Fig. 14:
Presenting parts of a building in section

The basic hatching for representing a material-independent interface of part of a building is drawn with an unbroken line at an angle of 45°. Narrow sections, such as profile cross-sections in steel construction, should be black for greater intelligibility.

LABELLING

Appropriate labelling is needed to produce a complete drawing, as well as lines, areas and hatching. The degree of detail in the labelling is chosen in relation to the scale; it is there to support the technical drawing (e.g. for stating dimensions, numbers of rooms, information about material etc.).

The chosen typeface must be unambiguously intelligible, so a standard typeface is usually chosen. Standard type (also called ISO type) is the name for an internationally used type of labelling using upper and lower case. Standard type is subdivided into four different forms on the basis of type sizes and type angles. Thus, a distinction is made between two type widths:

>Type form **A** – narrow, with a line width of height/14
>Type form **B** – medium wide, with a line width of height/10

and two type angles:

>Type angle **v** – vertical, with letters placed vertically to the direction of reading
>Type angle **i** – italic, with letters at 75° to the direction of reading

Using type styles

The best-known of the standard type forms used is a combination of type form **B** at the appropriate angle, producing **Bv** (medium wide line, vertical) and **Bi** (medium wide, italic).

Standard type forms are available as stencils for hand drawings in all the usual scales and type widths. Architectural drawings that are still produced by hand use building type. This is restricted only to upper-case letters developed from the shape of a square.

CAD programs can usually use the full range of fonts offered by the operating system. But here, too, it is better to choose a font in common use, as the next user should also have the font installed when data are exchanged.

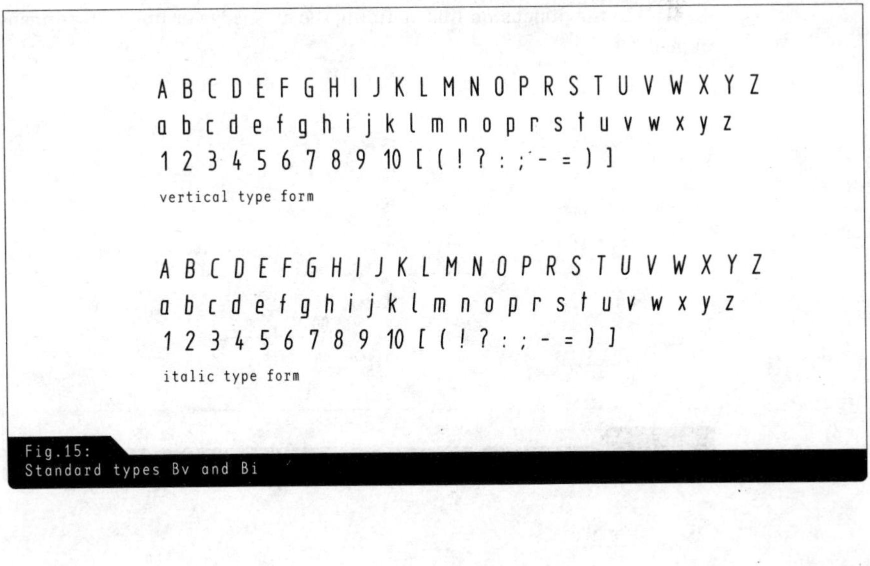

```
A B C D E F G H I J K L M N O P R S T U V W X Y Z
a b c d e f g h i j k l m n o p r s t u v w x y z
1 2 3 4 5 6 7 8 9 10 [ ( ! ? : ; ´ - = ) ]
vertical type form


A B C D E F G H I J K L M N O P R S T U V W X Y Z
a b c d e f g h i j k l m n o p r s t u v w x y z
1 2 3 4 5 6 7 8 9 10 [ ( ! ? : ; - = ) ]
italic type form
```

Fig.15:
Standard types Bv and Bi

Lettering is always positioned either horizontally to the direction in which the plan is to be read, or vertically turned anti-clockwise (and thus readable from below or from the right).

DIMENSIONING

Principles for dimensioning

Regardless of the fact that plans are drawn to scale, all the relevant dimensions must be clearly defined.

Regardless of whether plans are drawn true to scale, all the relevant dimensions must be defined clearly. This is done with the aid of dimension chains, relative elevations or by indicating specific dimensions. Dimension chains are sections arranged next to each other and provided with individual dimension information. Relative elevations are defined heights for particular points (for example the top edge of a floor slab).

Guidelines for entering dimensions on drawings are laid down in specific national standards (see appendix).

Dimension chains

Structure of a dimension chain

A dimension chain is made up of the following elements:

_ Dimension line
_ Auxiliary dimension line
_ Dimension limits
_ Dimension figure

Dimension limits

Dimension line, dimension limit lines and auxiliary dimension lines are always unbroken lines. The dimension line is placed parallel to the part of the structure to be dimensioned, with the auxiliary dimension line vertical to the dimension line, defining the axis, edge or line being dimensioned.

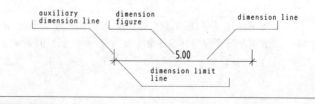

Fig.16:
Elements of a dimension chain

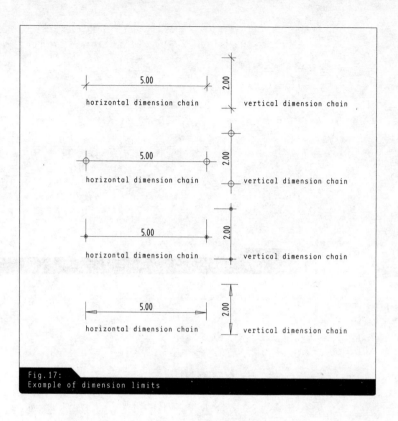

Fig. 17:
Example of dimension limits

Dimension limits define the external point of the dimensioned length on the dimension line. Even though in principle the auxiliary dimension line defines this, it is often unclear for intersecting lines on drawings whether the thick line for a wall through which a section has been taken also contains the dimension limit.

For this reason the dimensions should be clearly demarcated. According to the scale, either a diagonal line or a circle is used (e.g. design with lines, working plans with circles and detail planning with small dimensions chains with closed circles), but in principle a free choice can be made. Dimension limit lines are drawn at 45° from bottom left to top right seen from the direction of reading.

Dimension figures in dimension chains

The dimension figure equals the length of the structural element to be dimensioned and correspondingly also identifies the spacing for the dimension limits. For distances larger than 1 m the figure is given in the metre unit (e.g. 1), for distances smaller than 1 m the unit is the centimetre

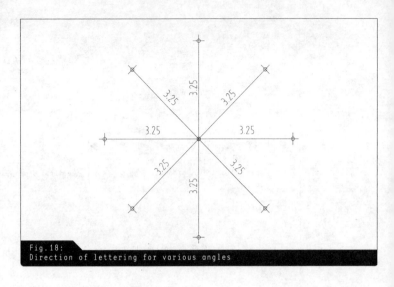

Fig. 18:
Direction of lettering for various angles

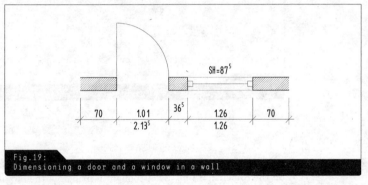

Fig. 19:
Dimensioning a door and a window in a wall

(e.g. 99 or 25). For millimetres the figures are raised (e.g. 1.25^5 or 36^5). Units such as m or cm are not given.

Position of the dimension figure

As a rule, the dimension figure is placed above the dimension line and centrally between the dimension limits. For aperture dimensions the height of the aperture is also shown under the dimension line. › see Fig. 19 If there is an additional sill (e.g. for windows), the sill height is given directly on the inside of the aperture (e.g. with SH or S = 75).

Dimensioning inside the dimension line is uncommon: as an alternative to dimension figures above the dimension line, the dimension figures may also be inscribed directly into the line, omitting that part of the line.

Problems occur when the distances between the dimension limits are very small, i.e. if there is little room for the figure (e.g. for lightweight walls and installations in front of walls). Here, the figure may be placed directly beside the dimension limit. ⟩ see Figs. 25–27 Dimension figures and dimension lines must not overlap, to maintain legibility.

Height dimensions

Height dimensions in elevations/sections

Height dimensions cover floor, sill and clearance heights and relate to a level ±0.00. This is normally set as the upper edge of the completed floor structure in the entrance area. The surveyor surveys and relates this point to mean sea level or zero level (ZL), so that the building is correctly positioned in terms of height. Hence the reference of ZL and ±0.00 should be defined on every plan. All height indicators then relate to the starting-point ±0.00 in terms of + and – signs.

A distinction should be made in the drawing between height dimensions of plan or top views and sections or elevations.

Height dimensions on plan views

Height dimensions in elevations and sections are given by height indicators: the graphic symbol for a height indicator is an equilateral triangle drawn directly with the height figure into the drawing of the structural element or onto an additional dimension limit line (e.g. outside the building).

For shell dimensions the triangle is usually solid black; for completion dimensions it is left as an outline triangle. Thus, the dimensions given can be clearly and correctly allocated even on small drawings.

Height dimensions on plan views and top views are also represented by triangles (solid for shell construction, outline for completion elements), but most commonly, circles with a line passing through them are used, with the completion dimension above the line and the shell dimension below it. Here, too, the dimensions can be directly identified by filling the semicircles appropriately. Another classification option is to use

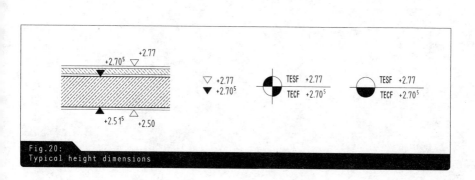

Fig. 20:
Typical height dimensions

the labelling TECF (top edge complete floor) and TESF (top edge shell floor).

Angle and curve dimensions

<small>Angle dimensions</small>

If structural elements are not at right angles to each other, the relevant angle must be indicated. This is usually done by stating the number of degrees and the typical symbol ∢, but it can also be done by using a segment of a circle with arrows on the ends and the number of degrees in the middle.

<small>Circular dimension</small>

For rounded structural elements, circular dimensions should be given, for example to define the developed length of a curved reinforced concrete wall. This is necessary, among other things, for determining and calculating the dimensions (continuous metres of wall, baseboard etc.). The dimension chain for a circular dimension consists of a circle parallel to the actual curve (i.e. dimension circle and curved wall have the same centre). The dimension limits can be used as explained above for circular dimensions, or arrows can be placed at the ends of the dimension circle.

Individual dimension indications

If individual dimensions must be indicated, they are usually written directly onto the particular structural element. They can also be allocated by abbreviations (e.g. SH for sill height), or symbols (e.g. Ø for diameter or □ for a rectangular section). Radii have a capital R before the dimension figure, screws and threaded rods an M. For simplification, height and width may also be given in abbreviations (e.g. W/H 12/16 for a wooden beam 12 cm wide and 16 cm high).

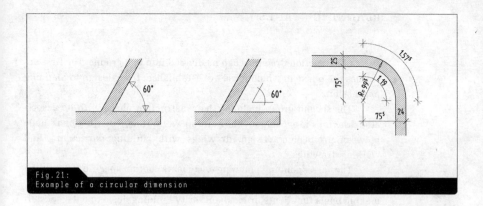

Fig. 21:
Example of a circular dimension

PLANNING STAGES

Construction drawings can be divided into two groups. The first describes the phase in which ideas are established: from design to planning permission.

The second group includes the construction phase with its associated drawings to accompany the building work. Thus, a distinction is made between pre-design, design, drawings with building particulars, and working drawings.

The corresponding planning documents contain specific information for a particular target group. Planning documents can provide a decision-making basis for clients or local authority building departments; specialist planners may use them as a basis for their own plans; and they can contain concrete building instructions for the specialist firms carrying out the work. The scope and precision of a plan derive from the purpose, nature and scale of a drawing. The less it is reduced, the larger the structural elements it shows are, and thus the dimensioning and labelling become more detailed.

DETERMINING BASICS

Land registry plans

Land registry plans (usually on a scale of 1:1000) of towns and municipalities are available to provide an overview or basis for a design; they may be slightly imprecise in their indications of dimensions.

As-built plans

If an existing building is being converted, its current state must be surveyed and presented in a drawing as a basis for the work. The subsequent planning phases are based on this. The degree of elaboration, and thus the precision, of as-built plans depend considerably on the use for which the building is intended. If a small extension is to be added to an existing home without particular demands on quality of detail, it is usually enough to record the relevant rooms in terms of width, length and height. If monument preservation measures have to be taken in a listed building, the dimensions must be recorded in detail and complemented with precise indications about surfaces and distinctive features.

PRELIMINARY DESIGN DRAWING

Purpose of preliminary design drawing

A concept is turned into an overall presentation in the form of a drawing (a plan) in the preliminary design drawings. A distinction is also made in this phase between construction drawings for future planning and concept drawings to explain an idea to a client. The aim of preliminary design planning is to clarify and explain the idea behind the design. Preliminary design plans thus express the planner's design approach and allow a great deal of latitude in terms of presentation. On the other

Fig.22:
Example of a land registry plan

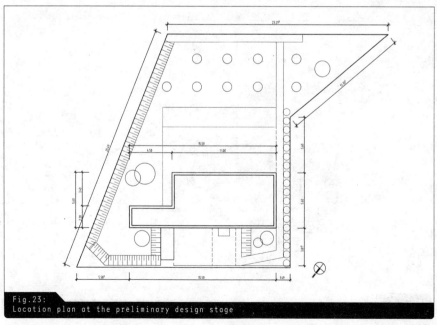

Fig.23:
Location plan at the preliminary design stage

hand, even a preliminary design plan may be consulted when clarifying information about the building with the relevant authorities. For this reason, it forms the basis for a preliminary decision by the building authorities.

Scale

Preliminary design drawings provide only the most necessary information about the shape and size of the building. These drawings are mainly drawn on a scale of 1:200, or 1:500 for large projects. Location plans are reduced even further (scales 1:500 or 1:1000).

Location plan

The location plan shows the building on the basis of the plot dimensions in the context of its surroundings. Its location on the plot is defined; hence the term location plan. The information entered gives a general view of the building's size and orientation, the nature of the terrain and its use, and where necessary includes the adjacent plots as well.

Drawing up plan views, sections and elevations

One sensible starting-point for drawing up further construction plans is a ground floor plan view matching the location plan. Starting with the ground floor plan view, the best way to develop the floors above is to base them on the plan view that has already been prepared (both at the drawing board and in a CAD program).

After drawing up the plan views, it is relatively simple to construct the sections. The chosen section line is first drawn into the plan view, and

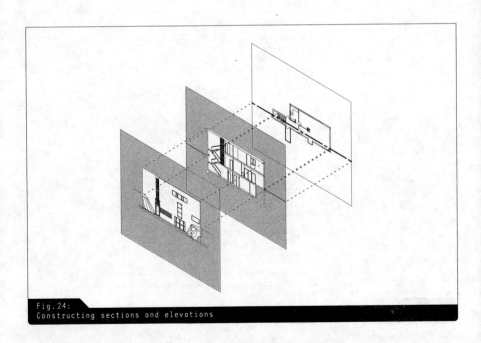

Fig. 24:
Constructing sections and elevations

then turned to be horizontal and form the basis for the section. It forms the zero line of the section, i.e. the entrance level for the building. All heights are developed vertically upwards on this line (upper floors) and downwards (basement floors): to do this, the edges of the intersected walls are simply extended to an appropriate level in relation to the zero line. Elevations should be constructed only on the basis of the heights marked in the sections. It is simplest to trace the external outlines of the sections and add windows, doors and ground connections by placing plan views underneath. ᐳ see Fig. 24

Representing structural elements

The aim of the preliminary planning is to clarify and illustrate the cubature of the building, the distribution of space and the way the building fits in with its surroundings, in terms of a provisional arrangement and provisional dimensions. Structural elements are usually presented without indication of materials, so that at first it is only possible to identify which elements are intersected.

In elevations, all the visible edges are represented by an unbroken line. The thickness of the unbroken line depends on scale, relevance (walls are more important than a door handle, for example) and the degree of detailing in the building. Thus, the outlines of the external walls and their apertures are emphasized most strongly.

Finishing details such as toilets, kitchen or furniture are given in top view and elevation, to illustrate the design. Furnishings are important to inexperienced clients as a yardstick for size ratios in housing construction in particular, so that they can understand proportions and the size of rooms.

Establishing scale

Representing trees, people and the external areas clarifies how the building fits into its surroundings while providing a background that does not conceal information. Proportions and scale can be conveyed better by showing objects whose size or proportions are familiar even to the unpractised eye. This is called establishing scale, and the individual items are scale-establishing objects. These "extras" are mostly used only at the preliminary and design stages, and for competition and presentation drawings.

Dimensioning

Dimensioning is restricted to rough measurements in preliminary design planning. In plan views, external dimensions and important room dimensions are given in order to make room sizes and overall measurements comprehensible. Individual projections and recesses, and door and window apertures, are not usually dimensioned.

In elevations, only important height dimensions such as eaves and ridge height are given; in sections the room or floor heights appear as well.

Labelling

Labelling is also restricted to simple identification of rooms by function and the estimated areas of the rooms in square metres.

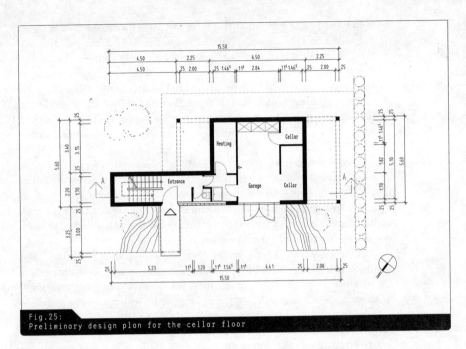

Fig.25:
Preliminary design plan for the cellar floor

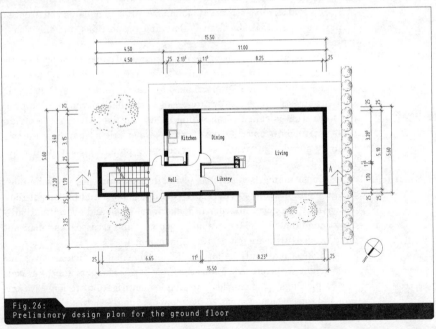

Fig.26:
Preliminary design plan for the ground floor

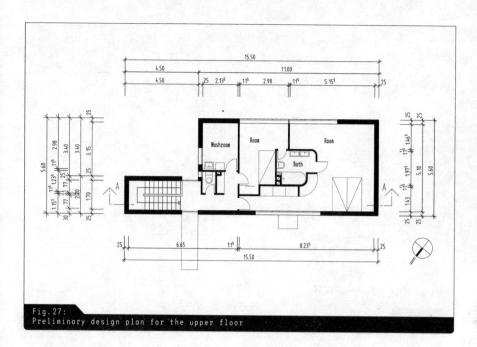

Fig. 27:
Preliminary design plan for the upper floor

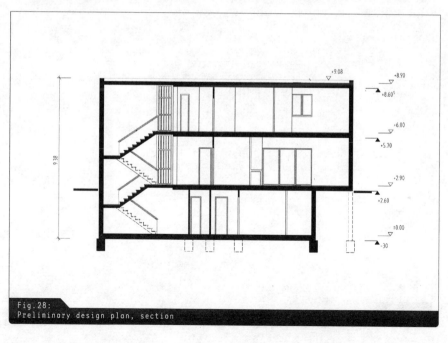

Fig. 28:
Preliminary design plan, section

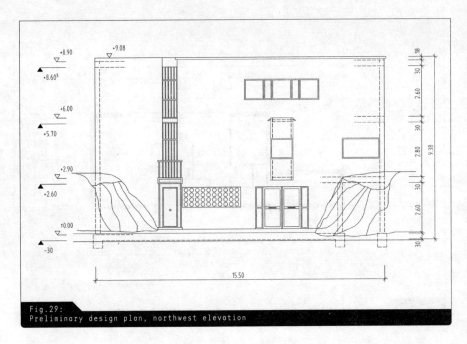
Fig.29:
Preliminary design plan, northwest elevation

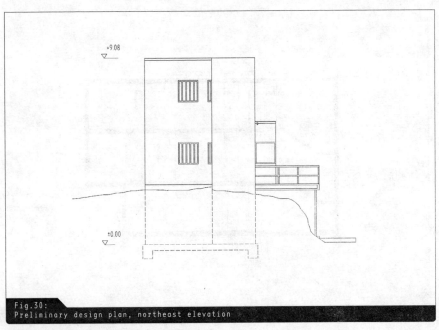
Fig.30:
Preliminary design plan, northeast elevation

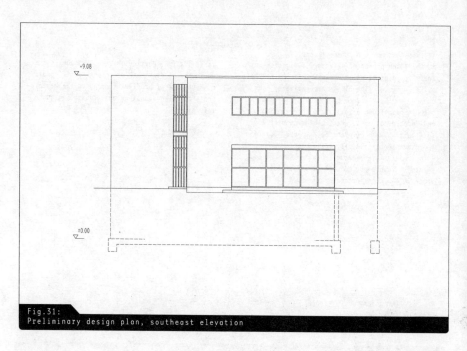

Fig.31:
Preliminary design plan, southeast elevation

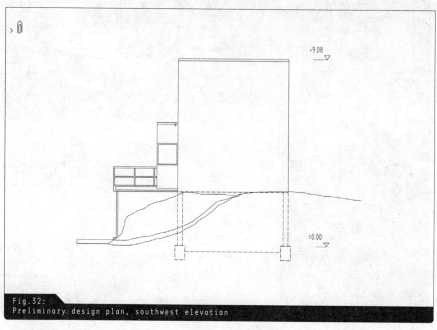

Fig.32:
Preliminary design plan, southwest elevation

\\Hint:
The building presented here is a design for an early residential building in Vaucresson (1922) by Le Corbusier. The plans were copied by the authors on the basis of Le Corbusier's plans, but dimensions and details have been changed and adapted, or measurement chains completed, to illustrate the relevant points here. Some of the dimensions for doors, toilets etc. are no longer admissible under current regulations and should therefore not be used as models for the reader's own construction drawings.

\\Tip:
It is helpful, particularly in the submission phase for student work or competitions, to draw a short list of headings summing up the key elements of the design idea and translating these into pictograms.

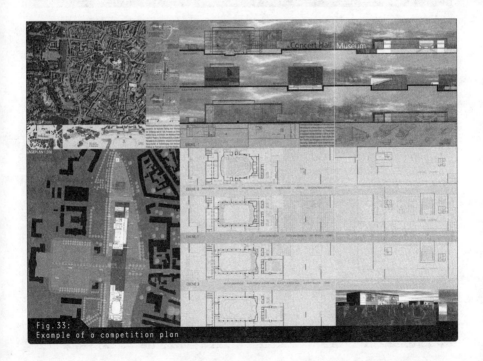

Fig. 33:
Example of a competition plan

PRESENTATION PLANS

Client presentation

To present a design, presentation plans are prepared independently of the classical construction drawings. Presentation plans are usually drawn up after a preliminary design has been completed, so that this can then be confirmed for further planning purposes: presentation plans are intended to persuade a particular target group of the design idea and the concept, so they should be devised with this purpose in mind.

If, for example, an inexperienced client is unable to associate a three-dimensional idea with a technical, two-dimensional drawing, then it makes sense to underpin his sense of space with three-dimensional representations or perspective views.

The client may have to present the design to a third party, and may need visual materials to help persuade them. These can be three-dimensional presentations or perspective views, or graphic presentations of zoning, pathway links, work areas or similar items on the basis of plan views and sections.

Architectural competitions

If planners or students take part in competitions, their own ideas must be presented in such a way that it makes a clear impact on the jury. As juries are usually made up of a mixture of experts and laypeople, the

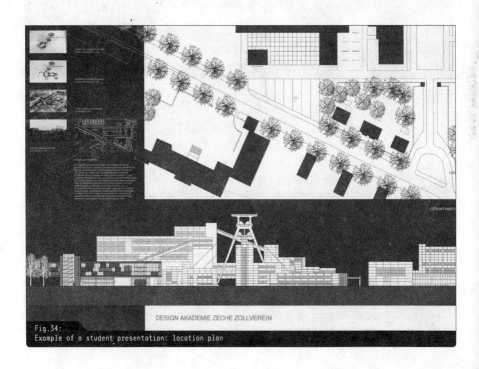

Fig. 34:
Example of a student presentation: location plan

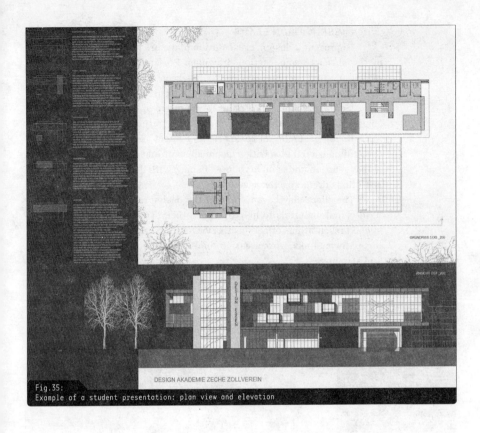

Fig.35:
Example of a student presentation: plan view and elevation

needs of both groups should be taken into consideration. As a rule there is little time available for judging the entries, and so everyone looking at the submissions needs to understand the individual designs quickly. It is also important to make your design stand out from those submitted by the other entrants.

Student presentations

When a student design is presented, the college or university lecturer responsible should be able to understand his or her students' ideas, i.e. the presentation should convince a qualified expert with appropriate skills in terms of abstraction and imagination. For this reason, more conceptual approaches tend to be chosen for student submissions than would be used for presenting ideas to a client. For example, if this is desirable in graphic terms, it is possible to omit the scale-establishing features that are needed for the layperson.

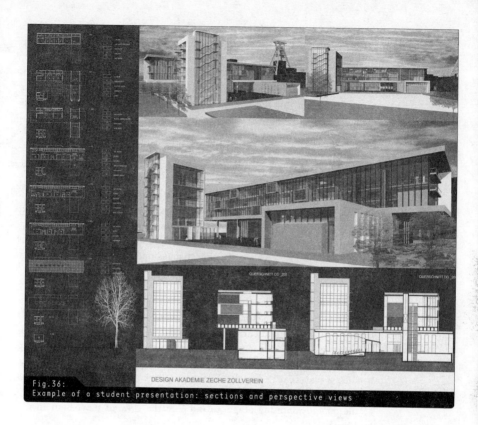

Fig. 36: Example of a student presentation: sections and perspective views

Content of presentation plans

As well as classical elements such as location plan, plan views, sections and elevations, presentation plans often contain three-dimensional representations of interiors and exteriors. Pictographic elements may also help to represent the design idea or the functional context of the design.

Devising presentation plans

The plans to be submitted are usually confined to one particular scale. For presentations, there is otherwise a greater degree of freedom than for technical construction drawings. Dimensioning can be reduced to a minimum, structural elements represented graphically or the whole scope of the drawing can be reduced essentially to elements of composition. – There are no limits on creativity. It is important to keep in mind the clarity and accessibility for the person looking at the work, and to devise the presentation appropriately.

DESIGN PLANNING

Purpose of design planning

Design planning is a further development of the preliminary planning stage. Now the architect and the client finally define the geometry and dimensions of the design for the building permission plans that will follow. The design plans must therefore show all the important elements that are relevant for the planning authorities' consideration. This phase also addresses planning in other fields like structural engineering and domestic services, which means that all the fundamental structural information (e.g. loadbearing walls) must be visible.

Scale

Design plan views for housing are usually drawn on a scale of 1:100, and for very large buildings on a scale of 1:200 or 1:500 if necessary. For example, if the plan view of a large industrial hall were presented on a scale of 1:100, a very large number of A0 sheets would be needed, so it would no longer be possible to take in the design as a whole. But, as explained above, since the design plans are intended to provide a comprehensible and useful basis for further discussion, it makes sense to choose a less refined presentation on a higher scale.

Showing walls

Even at the design planning stage, material-dependent hatching can be used for representing walls, in order to define the material, e.g. a reinforced concrete, masonry or dry construction wall. Line widths distinguish between loadbearing and non-loadbearing walls. It is not usual to show wall surfaces (e.g. interior rendering) at the design planning stage. All the doors and windows should be entered on the plan correctly, in such a way that the aperture dimensions and, where applicable, the sills can be seen. The direction in which doors open should also be shown at the design planning stage, to indicate the flow of movement in the building.

Foundations are indicated in consultation with the structural engineers, along with their construction details (individual foundations, ice wall, foundation strip), with their correct depths and widths. If it is necessary to explain the section, invisible parts can be shown as dashed lines.

Showing floors and ceilings

Sectional drawings of floors and ceilings are made up of the shell floor including hatching to show material qualities and the completed floor as an upper edge, in order to define the height of the structure.

Showing invisible elements

Since the horizontal section plane in a plan view lies 1.5 m above floor level (see above), structural elements placed above it are not visible in the plan. But to understand the geometry and the space it is often necessary to show these elements as well. They can be beams or girders, which divide a space into several section visually (the dimensions of the beams or girders are shown directly on the plan view drawing, e.g. B43/35); or stairs, whose upper run should be shown, with a junction point to under-

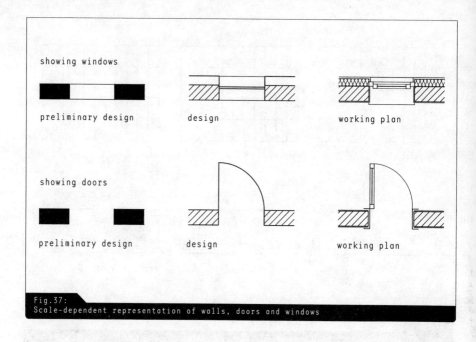

Fig.37:
Scale-dependent representation of walls, doors and windows

stand the geometry of the steps. The same applies to sections, for example in the case of covered mezzanine floors or invisible stairs. In elevations, the loadbearing walls and floors inside the building envelope can be indicated with a short dashed line.

Additional information in elevations

The following elements can also be shown in the elevations (views): windows with divisions and opening mode, blind cases BC), height of window apertures, balconies, sills, protrusions and recesses, roof forms.

Site layout

The existing and planned layout of the site before and after building should be drawn in as precisely as possible, as it is relevant for the building's entrances and exits, the necessary earthworks and the building inspection authorities. The building's height systems are also developed from this.

Showing stairs and ramps

Stairs are defined in design plans by stating the number of steps, risers and treads (e.g. 10 risers of 17.5/26). Flights of stairs are also identified by continuous lines, with the starting-point (starter) identified with a circle and the end-point (exit) with an arrow. Ramps are identified by two lines taken from the start of the ramp to the central point of its upper end.

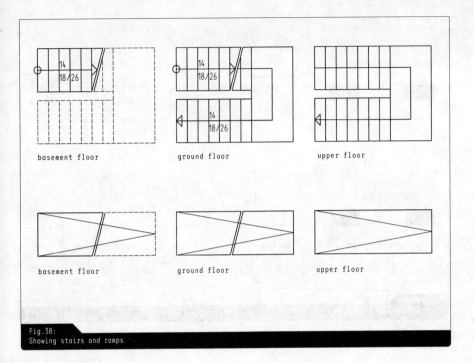

Fig. 38:
Showing stairs and ramps

In sections, the stair structure is shown as simply as possible, so that the geometry of the stairs can be understood: a distinction is made between concrete, closed or open steps in wood or metal as stairs with or without landing.

Dimensioning plan views

Dimensioning in design planning is intended to define the geometrical coherence of a building and the rooms it contains. The exterior dimensions of the building are entered first – as in preliminary design planning – including all external cladding and rendering. This makes it easier to determine the gross floor area, the gross space enclosed and the building's position in the location plan or on the plot.

> ◌

The second step dimensions all the external doors and windows, and ideally an additional dimension chain is added for the interior position of the windows. > see Fig. 39 Thus, all the apertures can be defined in their geometrical relationship with the elevation of the building and in relation to the spatial impact of the exterior wall. Any possible shifts between interior and exterior axes or possible window rabbets can be seen and planned.

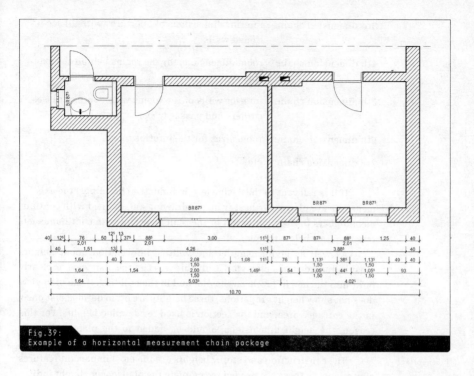

Fig.39:
Example of a horizontal measurement chain package

The next step records the interior spaces with the length and width. This is essential for calculating room and apartment sizes, and is useful for the later user as a basis for furnishing. It thus makes sense to give individual dimension chains for the overall room size and the wall features with door(s). The heights of apertures (doors, sills and windows) are indicated as described above with an additional figure under the dimension line, and sill heights are given for windows. Planners should pay particular attention to arrange dimension chains in a comprehensible order and on clearly distinguishable axes, so that the plan is easily intelligible. A typical sequence for a house works from the outside inwards:

1st dimension chain: total exterior dimensions (where applicable, additional dimension chain for walls with protrusions or recesses)

2nd dimension chain: exterior dimensions with all apertures (doors, windows, projections etc.)

45

3rd dimension chain: interior dimensions of apertures with all sectioned walls

4th dimension chain: room dimensions for the rooms behind the exterior wall

5th dimension chain: interior walls of the rooms with doors, recesses, corners and passageways

6th dimension chain: room sizes for the interior rooms

7th dimension chain: etc.

If it is relevant to the design (e.g. in industrial buildings) the axis dimensions are placed furthest from the drawing and defined with continuous numbers to the right or left of the plan view and with continuous letters above or below the plan view.

As well as giving lengths and widths, information about heights must be included in a plan view. Only this way can the plan view be clearly built into the height levels on the plot and in the building. If the plan view does not show height gradations, these heights are often defined just once, in the entrance area, and the section is used for detailed heights. For this reason, the height dimensions should be added to the plan views only when the section is complete.

Dimensioning sections

The construction and storey heights are the most important features of sections, as these are needed to complete the plan views. Height indicators are normally used for this › see chapter Principles of representation/dimensioning, supplemented by dimension chains. The height indicators show absolute

\\Hint:
Octametric dimensions with controlling and working dimensions should be taken into account for masonry buildings. A multiple of bricks 24 cm long including an A joint 1 cm wide, without summit joint, gives the working construction dimensions (11.5 cm; 24 cm; 36.5 cm; 49 cm;…), and with joint the specified dimension (12.5 cm; 25 cm; 37.5 cm; 50 cm; etc.).

Further information can be found in *Basics Masonry Construction* by Nils Kummer, Birkhäuser Publishers, Basel 2007.

\\Tip:
The structure of the dimension chains must be thought through carefully, so that not too much measurement information is defined twice, but also that nothing is forgotten. If there are a lot of small interior rooms, it is often tricky to accommodate all the dimensions in the dimension chain package. If necessary, if individual dimensions cannot be accommodated, an additional dimension chain can be placed directly inside the plan (within a plan view). This is generally more manageable than placing another dimension chain dealing with the whole building on the edge of the drawing, splitting up only a partial section of the measurement chain above it.

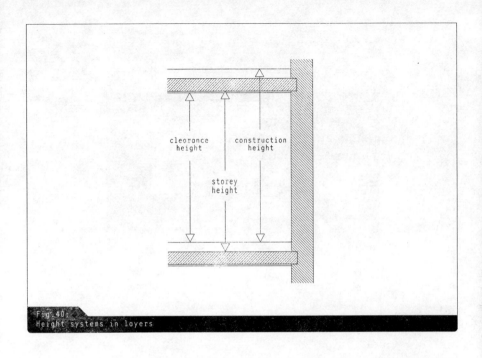

Fig. 40:
Height systems in layers

heights (related to zero level), and the dimension chains give individual construction and room heights. For example, height indicators in the roof structure give information about heights in the loadbearing structure and for the overall height of the roof structure (ridge height, attic height). When giving storey heights we distinguish:

- storey height: height from top edge to top edge of storeys in sequence
- clearance height (CH): height between top edge complete floor (TECF) and the lower edge of the complete floor above (where applicable, lower edge of the rendering or suspended ceiling)
- construction height: distance between the top edge of the shell floor (TESF) and the bottom edge of the shell floor above (BESF)

The following sequence should be followed when arranging the dimension chains (as in the plan view):

1st dimension chain: overall exterior dimensions (where applicable, additional dimension chain for recesses or protrusions)

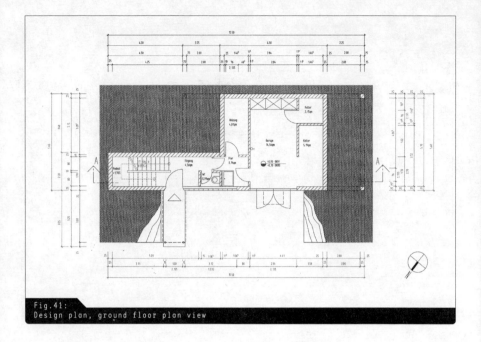

Fig.41:
Design plan, ground floor plan view

2nd dimension chain:	external dimensions with all apertures (doors, windows, sill height etc.)
3rd dimension chain:	interior dimensions with all apertures (doors, windows, sill heights etc.)
4th dimension chain:	clearance heights
5th dimension chain:	etc.

Dimensioning elevations

Elevations (views) are dimensioned with height indicators. It may be necessary to use dimension chains for completeness when dealing with larger buildings.

As for a section, the height indicators relate to the zero line at complete floor height on the ground floor, and additional information about height at zero level (ZL) can be appended.

Additional ZL information may be included, particularly to represent street height and the lie of the terrain.

Labelling

As a rule, only distinctive elements, for example the top edge of the land and the roof edge, are given in elevations at the design stage.

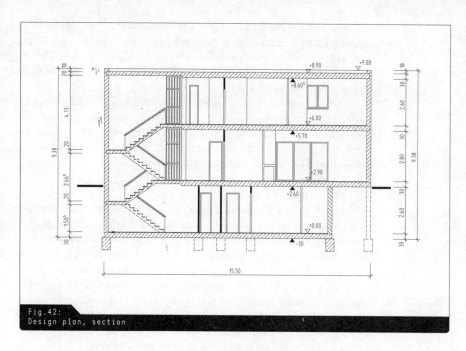
Fig.42:
Design plan, section

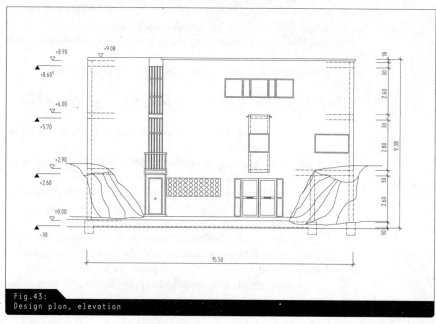
Fig.43:
Design plan, elevation

Additional information on the plan view drawing is given by room identifiers. These contain room numbers and/or room names (WC, living room etc.) and where applicable the room area in square metres. The entrance to the building is often identified by a solid black triangle, to make it easy to find on the plan.

Inclines such as rises and drops are identified by a height dimension and an angle, and in sections the percentage or degree is given, complemented by a directional arrow: e.g. roof pitch to the right, 45°. A north arrow is also drawn in, so that the light quality in the various rooms can be assessed, and so that the elevations can be identified.

PLANNING PERMISSION

In this phase, the location plan and the design drawings are completed by adding information in accordance with the regulations of the relevant planning authorities. Various requirements are imposed according to the nature and size of the planned building. In principle, it is very easy to turn designs plans into planning permission submissions.

Official location plan

Location plans, in particular, must often be drawn up by a publicly appointed surveyor or the local survey office. The responsible authorities should be asked to explain what is required before an application for panning permission is submitted. For an official location plan, written and drawn parts are distinguished. The location plan drawing is usually on a scale of 1:500, but scales 1:1000 or 1:250 are possible for very large or very small projects. The content of the location plan is usually in black and white, but colours may be used for areas and boundaries where applicable.

A location plan should contain the following information:

_ Location of the building plot relating to points of the compass, north arrow
_ Existing buildings identifying use, number of floors, roof shape (ridge direction)
_ The planned building identifying exterior dimensions, heights relating to zero level, number of storeys, roof shape
_ Exterior dimensions of existing buildings and new buildings
_ Information on type of use for the areas that are not built on, such as garden, parking space, playground, terrace etc.
_ Statement and verification of distances from adjacent plots and to public areas (often in a separate plan showing distances apart)
_ Marking and demarcation of areas with building restrictions
_ Position of supply lines (water, electricity, heat, radio/telephone)

\\Tip:
Converting from degrees to percent is simplified if the calculation is done in stages: a slope of x% gives a height difference of x cm over a horizontal distance of 100 cm. If a certain number of degrees is to be converted into a percentage, then the equation tangent angle = opposite leg/adjacent leg is used. For example, 10° gives the following equation: tan 10° = 0.1584. This corresponds to 15.84 cm to 100 cm, in other words 15.84%.

The location plan is supplemented by the following written information:

_ Scale
_ Details of street name and house number, owner, plot designation (boundaries, open fields, parcels)
_ Area dimensions, boundaries relating to land registry
_ Information about existing trees, especially if subject to nature conservation or tree protection orders
_ Information about areas with building restrictions and their use

If the building plot is part of a development plan its provisions must be complied with; these are usually shown by graphic symbols.

Building in existing stock

If the plan is not for a new building, but an extension or conversion, a clear distinction must be made between demolition and new construction in the plans. As a rule, as-completed plans are drawn at the beginning of the project, to plan conversion and demolition measures. If a great deal of the building is to be dismantled, it is also recommended that separate demolition plans be drawn up, to be developed further as working plans and used as a basis for the demolition work. Demolition is further identified by crosses at an angle of 45° and dashed in on the plan view and section. In addition, existing building, demolition and new construction are coloured on the plan:

Black: old, existing sections to be retained
Red: new sections to be added
Yellow: sections to be removed in the course of the building programme

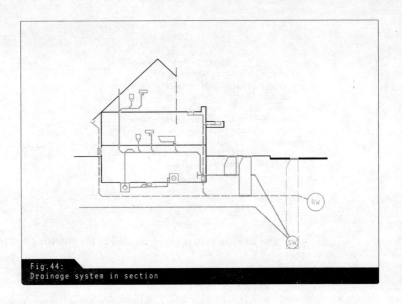

Fig. 44:
Drainage system in section

Generally speaking, the desired presentation should be discussed with the planning authority.

Drainage drawings

Drainage plans identifying the course of the drainage pipes are drawn to show drainage for sanitary facilities like WCs or kitchens and the exterior drainage of roof surfaces. They show the fall pipes and pipes connecting to the facilities in plan view and section. The pipes are identified by their diameter (e.g. DN 100 = interior diameter 100 mm), according to which wall-mounted installations, plumbing shafts etc. are dimensioned.

WORKING PLANS

Purpose of working plans

Working plans are intended to provide information that can be used to construct the building precisely. Hence, the working drawings must contain all the individual items of information needed to complete the building. These include the architects' plans, which should also be complemented by plans from other specialists, e.g. heating, sanitary pipework, loadbearing structure, fire protection etc.

The working drawings include two subgroups laying down the scale and degree of detail in the planning. The working plans are on the scale of 1:50 and the detailed plans on scales of 1:20 to 1:1. Fundamentally a distinction can be made between the following working plans – not necessarily a complete list:

_ Working plans scale 1:50: these consist of plan views, elevations and sections representing the building as a whole or parts of it.
_ Façade sections scale 1:50–1:10: as a rule, the façades are shown in more detail with sections, interior and exterior views, in order to fix the relationship with other parts of the building in structural and geometrical terms.
_ Installation drawings scale 1:50–1:20: special installation plans are drawn for individual services. These include screed plans, tiled surfaces, floor covering plans, plans for dry construction and grid ceilings etc.
_ Detail plans scale 1:20–1:1: detail plans show individual structural points or connection with every element, in detail.
_ Site installation plan, where applicable

Working plan presentation

The planner must translate the design planning into valid and complete working, detail and construction drawings in such a way that the contractor doing the work can understand and implement them without difficulty. The working planning must be so precise that no unintended scope for interpretation or dimensioning is allowed. But it does not have to define every screw in detail, as it can be assumed that the construction specialist will have the necessary expert knowledge. Possible requirements laid down by the planning authorities are also built into the working plans. Working planning must be continued and adapted during the building phase, if changes or confusions arise.

Representing intersected structural elements

As a rule, the intersected structural elements are presented on a scale of 1:50 so that they can contain direct structural statements about the wall structure (e.g. masonry rendered on both sides or reinforced concrete ceiling with floating screed). Openings and slits in walls and ceilings must also be shown. These are often needed for service installations (e.g. fireplace, ducts for heating, sanitary, ventilation and electrical services) and should be agreed with the planner responsible. Slits are shown by a diagonal line; as soon as the structural element is cut through completely, the opening is crossed through. Triangles can be solid black for emphasis.

Axes in working plans

For buildings with repeating loadbearing systems, such as industrial or office buildings, the longitudinal and lateral axes of the loadbearing frame should be identified by numbers or letters. Thus, individual areas can be categorized directly in such a way that the structural engineer can relate to the system and work with the axis definitions. Axes are drawn with dash-dot lines, either throughout the drawing or outside the building only.

Dimensioning working plans

All the dimensions needed for correct execution of the plans must be included at this stage. They include all the dimension chains already

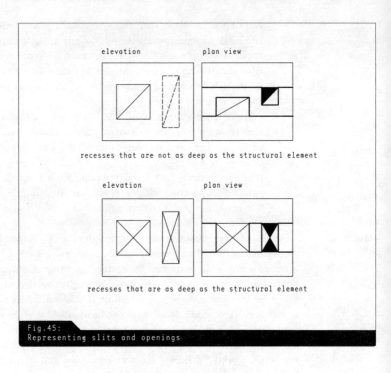

Fig.45:
Representing slits and openings

> ✎

shown in the design plans; but also the height, width and depth of every relevant structural element, without any omissions, and each element must be attached to a part of the building that can be used for on-site measuring. First, the general dimension chains are placed outside the building in the design plans; the detailed measurement chains are given inside the plan view or section.

Room labels

The room labels contain much more information in the working plan than in the design plan. In addition to room numbers and room designations, we include:

_ the room area (A) in square metres
_ information about the surrounding walls (S), for example measurements of skirting boards
_ clearance height (CH), which is needed in turn for wall developments, e.g. measuring up for painting

It is generally also possible to specify floor, wall and ceiling fittings. As they cannot usually be accommodated inside a room, the fittings can be

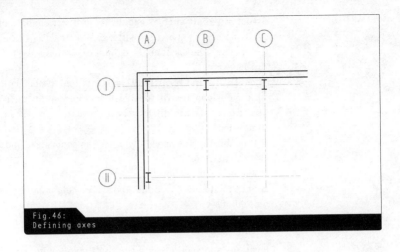

Fig.46:
Defining axes

named in the plan header › see chapter Plan presentation and allocated to the rooms by means of abbreviations (e.g. W1, W2 for wall fitting etc.).

Working drawings

Content of plan views

Plan views should give the following information:

_ The nature, qualities and dimensions of structural elements
_ Material-related presentation of wall, ceiling and floor fittings
_ Representation of the sealing and insulation levels
_ Door and window apertures with direction of opening, aperture and sill heights
_ Stairs and ramps with walking line, details on number and relation of steps and oversteps where applicable
_ Structural element qualities such as fire protection and sound insulation

\\Tip:
When dimensioning, you should imagine yourself in the role of the construction specialist carrying out the work on the building site, who has to implement the plans on the spot. For example, if a door is to be set with its outer edge flush with the jamb, it makes little sense to dimension the centre axis of the door. Double dimensioning should also be avoided as far as possible, because it increases the amount of work needed to make a change made, and double dimension chains are easily overlooked, which can lead to inconsistencies within the plan.

- Structural joints such as expansion joints or changes of covering
- Wall and ceiling apertures, slits, shafts etc.
- Technical fittings, channels, chimneys, drainage systems, underdrainage etc.
- Fixed fitting and furnishings, sanitary and kitchen equipment
- All dimensions of structural elements needed for the correct construction of the building (every protrusion or recess must have a dimension)
- All the dimensions needed to establish room sizes and for quantity calculations
- Room labels (see above)
- Height related to ZL, so that the storey height can be assigned unambiguously
- Detail references

Content of elevations and sections

Elevations and sections will contain the following additional information:

- Storey heights, clearance heights, shell heights
- Height markers for shell and finished floor, foundations, roof edges etc.
- Floor and roof structures
- Presentation of the existing and planned organization of the terrain surface
- Windows and doors with graphic representation of divisions and modes of opening
- Gutters, downpipes, chimneys, roof structures
- Covered intermediate floors, loadbearing walls and foundations as dashed lines
- Course of foundation pit drawn in
- Structural ceiling, roof and floor details
- Glass specifications in elevations, if different

Façade section

It can make sense to draw a façade section in order to show the entire façade in detail and not take out individual points of detail. This will show the complete height development in section, interior view, exterior view and where applicable complemented with a plan view detail with all connection points and height relations between interior and façade.

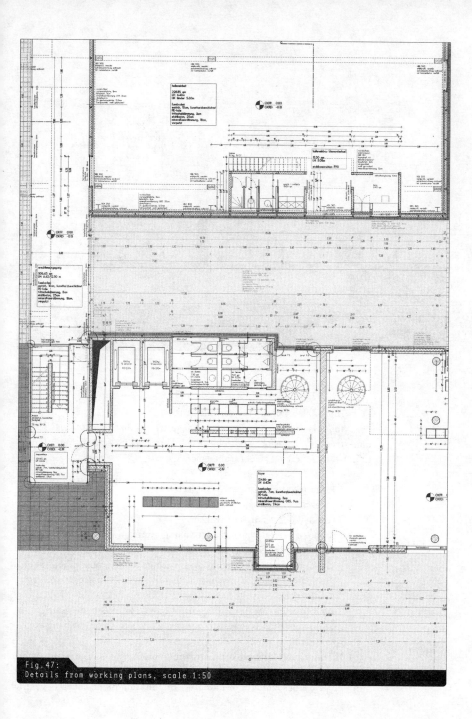

Fig. 47:
Details from working plans, scale 1:50

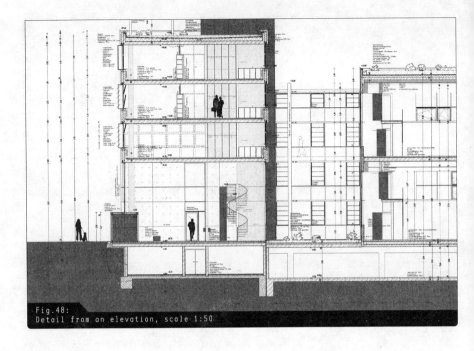

Fig.48:
Detail from an elevation, scale 1:50

Installation plans for service areas

Installation plans are construction drawings that show the way particular structural elements are to be fitted. They may include fitting information for the following structural elements:

_ Prefabricated reinforced concrete parts
_ Steel parts
_ Wooden beams or roof trusses
_ Areas of screed (showing expansion joints and conduits)
_ Laying stones (with layout grid, intersections and expansion joints)
_ Tiled surfaces (with layout grid, fittings and expansion joints)
_ Suspended ceilings (layout grids, fittings, acoustically effective areas etc.)
_ Double or cavity floors (layout grid, fittings under the floor)
_ Floor covering (layout grid or axes, change of floor covering etc.)

Installation plans are often drawn up on the basis of existing working plans, with the appropriate additional information entered, e.g. through grid lines, colours or hatching. Installation plans cover a specific

Fig.49:
Example of a façade section, scale 1:20

service area and are usually drawn up before tenders are invited for the particular field, so that the plans can be enclosed with the tender information.

Detailed plans

Detailed plans include all sorts of connections, system structures and transitions. As well as fixing control equipment in the drawing, there are particularly important points at which various control systems coincide or merge with each other. It is impossible to make a general statement about the detailed plans needed for a project, as they depend strongly on the individual project, the depth of detail required, the planning demands, and questions and uncertainties within the firms doing the work. Typical areas where detail is needed are:

- Façade: window joints and systems, transition between ground and rising walls, connections between façade and roof, corner situations, external doors, balconies, sills and parapets, shades against sun and glare
- Footings: foundations, drainage, sealing, insulation at ground level
- Roof: attic, eaves, ridge, verge, gable, roof apertures such as chimneys, ventilators, skylights and roof windows
- Stairs: system section, upper and lower connection, landings, banisters, handrails
- Floor and ceiling fittings: system sketches for all ceiling fittings used, transitions between different floor types, connections to rising building sections, fixtures, conduits
- Doors: system doors, frame systems, steel-framed doors, lift doors, shaft caps
- Dry construction: connections of walls to façade, shell structure, floor and ceiling, suspended ceilings
- WCs, kitchens, fitted furniture: structural details, connections, WC dividing walls etc.

Site installation plans

Site installation and organization plans coordinate the site and the construction firms involved. It is often unnecessary to draw up a separate site installation plan for small projects. But if there is only a small amount of room on site, it makes sense to do so, to avoid getting in each other's way and using the plot inefficiently. Hence, the following should be recorded in a site installation plan:

- Storage areas and working areas/site road
- Site management containers

_ Accommodation and sanitary facilities
_ Working areas around the building
_ Excavation pit
_ Lifting equipment (e.g. cranes) with radius and operational area
_ Building fences, entrances, signs etc.
_ Areas for individual trades (e.g. bending and cutting areas for concrete construction)
_ Soil storage areas where applicable
_ Power and water supplies, disposal facilities, rubbish management etc.

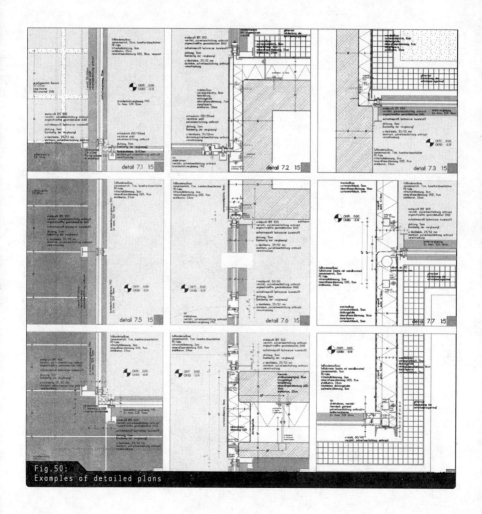

Fig.50:
Examples of detailed plans

Workshop drawings by firms involved in the project

Construction firms in different service areas involved in the project make their own workshop drawings on the basis of the working plans (sometimes also called works drawings). These are submitted to the planners before building starts and must be examined and signed off by them. Typical specialist firms who make their own workshop drawings include the following service fields:

- Metal or steel construction work (windows, steel constructions, railings etc.)
- Timber or carpentry work (wooden structures, roof trusses, windows etc.)
- Ventilation construction work
- Lift construction work

SPECIALIST PLANNING

Loadbearing structures

Working plans for structural engineers

Structural engineers draw their own working plans, placing particular emphasis on statically relevant elements. Which plans are drawn depends primarily on the choice of building materials. If the building is to be in reinforced concrete, encasement and reinforcement plans should be prepared; for wood or steel construction, plans for the appropriate rafters, timber and steel construction.

Position plan

› 🖉 › 🔎

Position plans show individual positions to clarify the statical calculations. The positions are numbered on the basis of the design drawings, and these numbers are also found in the statical calculation.

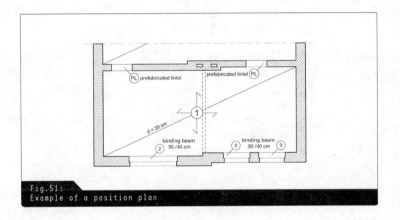

Fig.51:
Example of a position plan

🖉

\\Hint:
Plan views for structural engineering planning do not usually present a top view of the floor below, but of the ceiling above it. A structural engineer's plan view for the 2nd floor shows, as well as the identically intersected structural elements, a different view from the architects' plans. For clarification you should imagine a mirror-floor showing all the outlines of the ceiling.

🔎

\\Example:
A reinforced concrete beam is calculated statically and allotted the number 21. This beam is then also given the number 21 on the position plan, to make it clear to which beam in the building the statical calculation applies. If applicable, all the beams that are the same can be allotted the same number.

Encasement and reinforcement plan

Encasement and reinforcement plans are drawn up for reinforced concrete buildings. Here, shell plans show the structural elements to be encased (e.g. a reinforced concrete ceiling or wall). Encasement plans are particularly important if the subsequent surface is to be important visually (e.g. for fair-face concrete walls).

Encasement plans show the ceiling above the storey shown with:

_ axis, mass and height
_ supported or loadbearing construction elements
_ cavities relevant to loadbearing
_ types and strength classes
_ directions of span

Reinforcement plans contain information about reinforcement mats and bars to be built into a reinforced concrete section. A mat is usually shown as a rectangular area with a diagonal line marked with the mat type. Additional information for reinforcement plans:

_ Concrete steel types
_ Number, diameter, shape and length of steel bars and welds
_ Concrete strength classes, concrete covering
_ Apertures and special structures
_ Precise steel or item lists for construction to complement the drawing

Timber construction plan

>

Timber construction plans are drawn up for wooden structures, showing the precise position and dimensions of the individual timber construction elements (beams, supports, purlins etc.). As a rule, axes are dimensioned and connection points shown separately in detail. For example, if a pitched roof is to be built, the position and dimensions of the purlins and rafters must be shown in a rafter plan.

\\Hint:
Timber constructions and rafter plans are described in detail in: *Basics Roof Construction* by Tanja Brotrück, published by Birkhäuser Publishers, Basel 2007.

Building services

Special plan are also drawn up for building or domestic service installations, providing a basis for the installing the service equipment. Separate plans are usually made for each service. Particular services include:

_ heating installation
_ water supply and sewerage installations
_ ventilation installation
_ electrical installations
_ fire technology and alarms
_ data technology
_ lift technology

As well as the actual rooms for services, such as a utilities room, boiler room or similar, the pipe runs, holes and cable runs are the most important features. A plan showing slits and holes is often prepared to be included in the architects' plans, giving precise details of interventions in the shell.

PLAN PRESENTATION

If a drawing is to be issued in paper form, the paper format must be established to meet the demands of modern reproduction, and given a plan header.

PLAN COMPOSITION

Drawing area

Once the size of the building or section of a building to be shown and the scale of the drawing are fixed, the necessary amount of drawing space can be calculated. Then an appropriate area is added to the building dimensions on both sides to allow space for the dimension chains, and the drawing area needed is calculated on the basis of the scale. A plan header must be accommodated next to the drawing area (see below) and where applicable a frame, allowing for the intersecting edges.

Choice of paper format

All kinds of plan formats with different side measurements are possible for presentation purposes (competitions, student assignments). For example, a long, narrow building could be presented on paper with the same proportions, which would enhance the effect of the building's shape.

It generally makes sense to choose a common paper format for construction drawings (e.g. the DIN A series), as this can be reproduced easily. Large-format paper is needed for working plans, while for detailed drawings it is usually advantageous to use a format such as DIN-A3, which can be duplicated on most photocopiers. ❯ see chapter Principles of representation

Different scales

Drawings can also appear on different scales on a single plan: for example, it can make sense to show appropriate details of individual anchor points alongside a façade section. Care should be taken so that the scale for the individual drawing can be identified precisely through clear labelling.

PLAN HEADER

Each plan contains a plan header, showing clearly which project and what details this particular plan shows. The header is usually placed in the bottom right-hand corner of the plan. Plan headers for presentation and construction drawings are organized differently.

Presentation plan header

If the plans are intended for presentation purposes, they should have a corresponding plan header. The overall graphic design of the plan usually includes the header. As well as the project name, it gives details of scale, plan contents, (e.g. ground floor) and author. Supporting design pictograms, explanatory sketches on the section, or plan view level and north arrows can also be included in plan headers.

For university and college projects, the student's matriculation number should be included, as well as his or her name, department or

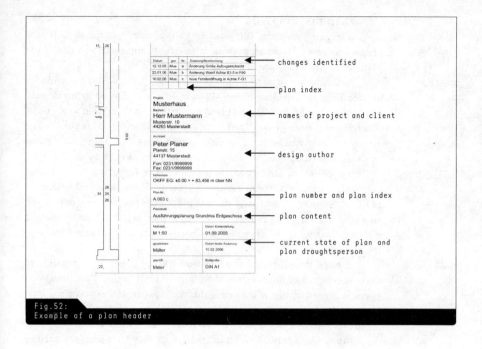

Fig.52:
Example of a plan header

Construction drawing, plan header

professor. For competitions, names or attributions do not usually feature on the plans. Instead, a camouflage number is often provided in the competition documents, and is placed in a particular corner of the plan. It is submitted with the competition plans in a sealed envelope, which will be opened only after the competition has been judged, to reveal the author of the design.

For construction drawings, the plan header first names the client, the author of the design, the plan draughtsperson and the scale. The current status of the plan is indicated. As new developments or specialist planning material have to be worked in regularly, the precise date of the current plan must be recorded. It is also a good idea to give the dates of the first version and the current revision. It is also helpful for understanding the additional data to include a summary of the plan changes to date. Building permission plans usually have plan numbers, which define them unambiguously. For simplicity, the current position expressed by the plan can be incorporated in the plan number: Plan A34d then means: working plan no. 34, fourth revision (d). Of course plan numbers can be allotted as wished; it is easier to administer the plans if a system is established at the beginning of the project and agreed with all concerned.

PLAN DISTRIBUTION

It is rare nowadays for plans to be drawn by hand, and CAD programs are generally used, making it easier to duplicate and distribute plans than when they are drawn by hand.

Plotting

CAD files are printed on plotters; smaller formats can be produced on standard commercial printers. Plotters are large-format printers, usually in the A1 or A0 format, and plans are printed on them from paper rolls. The roll widths are generally 61.5 cm for A1 rolls and 91.5 cm for A0. The size of the plan can be input individually in most CAD programs, so that any plan format compatible with the width of the roll can be used.

Copying

If a plan that already exists on paper is to be copied, large-format copiers are available. These usually operate with the same roll widths as plotters. Colour copies of large-format plans are generally very expensive, so a plan should be reproduced by multiple plotting. It is worth choosing a plan format for detail drawings that can be duplicated on standard commercial copiers.

Blueprints

If hand drawings are produced on tracing paper, it is possible to make blueprints of them, rather than large-format copies. Here, blueprint paper is placed on top of the original tracing paper, and exposed to ultraviolet light in a blueprint machine. Blueprint papers exist in various colour shades, and can then be called black-, red- or blueprints.

Blueprints are rarely used today because they have been replaced by CAD drawings, but they offer a reasonably priced possibility for reproducing hand drawings.

\\Tip:
Copy-shops often have mechanical folding machines that make it easier to fold a lot of plans. If you have to fold the plans yourself, it is best to use a sheet of A4 paper and a set square. First, take the left-hand side of the plan and fold it upwards to the width of the A4 sheet. The fold can be smoothed down with the help of the set square. Then the right-hand side is folded to a width of about 19 cm, until the remainder can be folded in the middle if necessary. Now fold the remaining strip back to the height of the A4 sheet (several times for larger formats). The inner corner, as shown on the diagram, is then folded inwards, so that is not perforated at the subsequent hole-punching stage. If you use the plan header and plan frame frequently on a CAD plotter, it is possible to draw in little lines on the plan frame, so that there is no need to measure, and the plan can be folded directly along the lines.

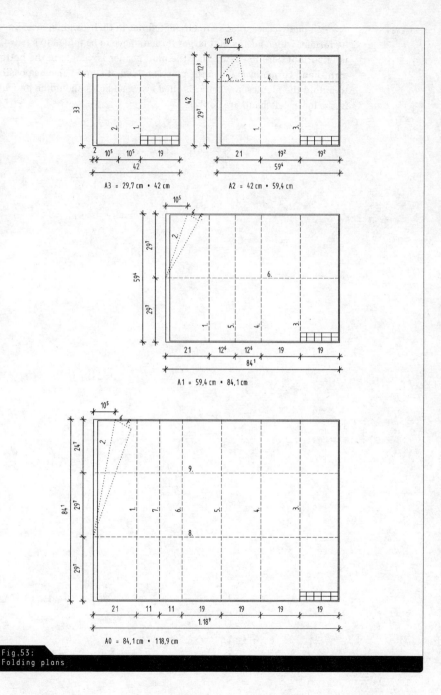

Fig.53: Folding plans

Folding

In many cases, plans are circulated not in their original size, but in A4 format, which means that larger formats have to be folded to A4 size. It is important here to ensure that the plan header is visible in the bottom right-hand corner even when the plan is folded. It should also be possible to unfold the plan even when it is filed away, so holes should be punched in the lower left-hand area only.

APPENDIX

SYMBOLS

Table 4: Symbols for sanitary fittings	
Item	**Symbol**
Hand basin	
Basin	
WC with cistern	
WC without cistern	
Bidet	
Urinal	
Shower-base	
Bathtub	
Washing machine	
Tumble dryer	
Corner bathtub	

Table 5:
Symbols for kitchen fittings

Item	Symbol
Cupboard under	
Cupboard over	
Cupboard under and over	
Sink with draining board	
Gas cooker	
Electric cooker	
Cooker with oven	
Built-in oven	
Worktop	
Refrigerator	KS
Freezer	
Dishwasher	GW
Microwave oven	
Extractor hood	
Chair	
Table	

Table 6:
Symbols for furniture

Item	Symbol
Cupboard	
Armchair	
Sofa	
Table	
Chair	
Grand piano	
Upright piano	
Desk	
Wardrobe	
Single bed	
Single bed with bedside table	
Double bed	
Double bed with bedside table	

STANDARDS

Large areas of Technical Drawing are covered by international standards. For this reason, they are standardized by ISO standards, which are recognized in most countries. The following table gives the valid ISO standards, which have generally been adopted by the national standardization institutions (ISO 128 would then be called DIN ISO 128 in Germany, for example).

Table 7:
Relevant ISO standards

ISO standards	Description
ISO 128	Technical drawings. General principles of presentation
ISO 216	Writing paper and certain classes of printed matter. Trimmed sizes. A and B series
ISO 2594	Building drawings – Projection methods
ISO 3766	Construction drawings. Simplified representation of concrete reinforcement
ISO 4067	Building and civil engineering drawings – Installations
ISO 4069	Building and civil engineering drawings – Representation of areas on sections and views – General principles
ISO 4157	Construction drawings – Designation systems
ISO 5455	Technical drawings. Scales
ISO 5456	Technical drawings. Projection methods
ISO 6284	Construction drawings – Indication of limit deviations
ISO 7518	Construction drawings. Simplified representation of demolition and rebuilding
ISO 7519	Construction drawings. General principles of presentation for general arrangement and assembly drawings
ISO 8048	Technical drawings. Construction drawings. Representation of views, sections and cuts
ISO 8560	Construction drawings. Representation of modular sizes, lines and grids
ISO 9431	Construction drawings. Spaces for drawing and for text, and title blocks on drawing sheets
ISO 10209	Technical product documentation. Vocabulary. Terms relating to technical drawings
ISO 11091	Construction drawings. Landscape drawing practice

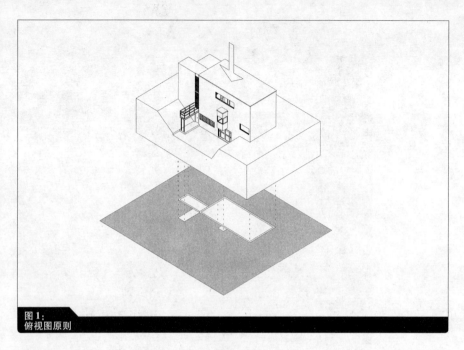

图1：
俯视图原则

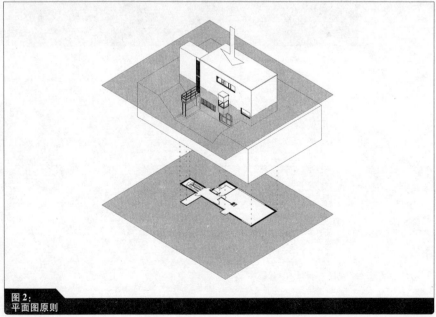

图2：
平面图原则

投影分类

建筑是通过设计草图和自由透视图来表示的,但是也使用各种其他的结构制图方式。在表示建筑物的外部信息时,应区分俯视图和正面图;在表示建筑物的内部信息时,应区分平面图和剖面图。

俯视图 (屋顶平面图)

俯视图制图是从上部俯瞰的视图或者投影图。俯视图(也通常称为屋顶平面图)对于总平面图是很重要的,例如,这可以在测绘上对建筑物进行定位。

平面图

同样,平面图显示了建筑物中某一具体楼层的信息。在平面图中,剖切面取为距楼面以上大约 1~1.5m 高度的位置,以尽可能多地包括砌体墙中的洞口(门、窗)。为了使图纸易于理解,应给出与结构剖面图相关的所有高度(窗台高度、洞口高度、楼面标高、楼层高度)和所有相关的平面尺寸。为了描述尽可能多的设计特征,以及表示出任意的窗户,如标高更高一些的窗户(窗户的不同位置可以通过给出窗台高度来区分)可以通过改变水平剖面的高度来实现。就平面图的视图方向而言,有两种基本可能性:

— 建筑学设计中用到的下行平面视图(俯视图),采用这种视图,可以记录房间的结构、形式以及尺寸;
— 自下而上的反镜平面视图,采用这种视图,可以表示出在水平剖面以上的建筑元素。结构工程师更喜欢这种视图方式,因为它能够表示出顶棚以上的承重结构。(参见"专业图纸"一节)

平面图的命名

平面图通常都是依据他们所表示的楼层来命名的,如地下室平面图、地面层平面图、第二层平面图、屋顶平面图等。在设计中,如果不能很明确地给楼层命名,例如楼层标高有高差时,显然需要用特殊平面来命名了,如第三层平面图、地下停车场平面图。

> **提示:**
> 作为惯例,俯视图和平面图都是北向的,换言之,设计图的上边缘表示北向,并且用指北针来表示。

立面图

立面图（在 ISO 标准中也称为视图）表示了建筑物的外表及所有的洞口情况。建筑的容积视图所表现的信息是关于建筑物的环境、形式和比例、类型及可适用的材料质量之间的关系。立面图与平面图、剖面图一起构成了一个完整的设计。

立面图是建筑物立面的平行投影图，即从建筑物的一侧看去所形成的投影图。因为投影线以一个合适的角度与投影面相交，所以偏移的剖面图将不能表示正确的尺寸。

立面图通常表示了建筑物的直接周围环境。这种周围环境包括了地形的走势，及与已有的建筑物协调适应。

立面图的命名

立面图是通过所处的方向来命名的。总平面图和平面图上的指北针确定了建筑的方向。因此，四个立面图的命名如下：北立面图、南立面图、东立面图和西立面图（或者东北立面图、西南立面图等）。如果只有两个立面图是可见的（如阶梯式住宅），也可以通过建筑物在图上的位置来命名，或者由其在整体规划中的位置来命名。但是，这就意味着只能明确固定两个侧面，例如花园或院子一侧，街道的一

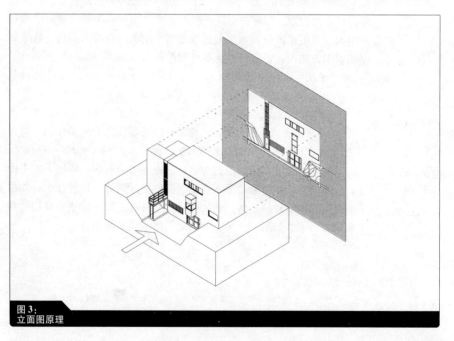

图 3：
立面图原理

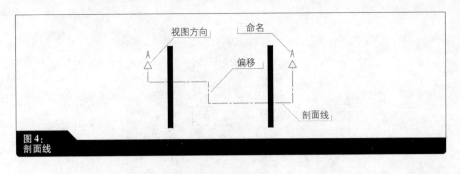

图4：
剖面线

侧。立面图的标记必须是十分清晰的，即使那些对建筑物所处地域并不熟悉的人也能一目了然。

P10

剖面图

剖面图是在建筑上做一个垂直的剖面，并且把这个剖面当作平行投影面。剖面图给出了关于楼层高度、材料质量以及所用建筑材料等方面的信息。

剖面线必须出现在平面图中。剖面线由粗点画线和视图的方向来确定，通过箭头和两个同样字号的大写字母来给定剖面图的方向和名称。用这种方式取出的剖面图可以记录与建筑物相关的所有信息，这也意味着剖面图有时可以偏移。剖面线的偏移必须是一个合适的角度，并且在平面图上标识出来。

剖面图元素

在一个剖面图上显示的重要元素包括屋顶、楼板和顶棚、基础，及带有洞口的墙体等。剖面图还应该表示出通过楼梯、电梯、坡道等建筑物的入口。

剖面图的
命名

通过平行于建筑物主轴取出的剖面图成为纵剖面图和横剖面图。纵剖面图是沿着建筑物的长边方向剖切得到，而横向剖面图则是沿着短边方向剖切得到。如果有两个以上的剖面图，则通常用大写字母或者数字来命名。由于在平面图上剖面线是用相同的字母在两端来标识的，因此剖面图的名称也相应地叫做 A–A 剖面图、B–B 剖面图、C–C 剖面图等。

P11

三维视图

轴测图

轴测图是在第三个平面上增加了高程的平面图或者视图。在规划阶段经常采用三维视角的轴测图，用以展示建筑物的空间效果。轴测图只是在特殊的情况下才被用作施工图，如角落的设计。

从二维图很容易得到三维图。但是，需要对投影角度加以区分

79

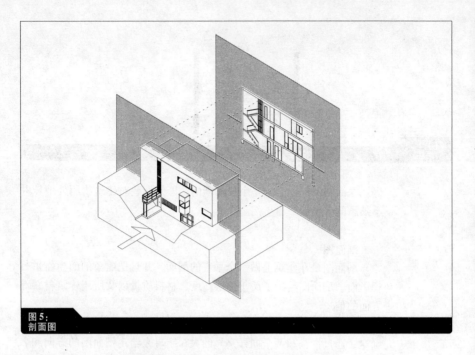

图5：
剖面图

（即使"投影"一词在这里有些令人误解）：

——"军事投影"：平面图绕某一顶角旋转45°，并在垂直方向增加高程；

——"建筑投影"：平面图绕某一顶角旋转30°或60°；

——"斜等轴投影"：立面图或者剖面图通过一条夹角为45°的附加线而得到的三维图。

等轴测图和正二轴测图

通过移动合适的角度得到基线，可以生成更好的三维视图。这正是等轴测图和正二测图的功能所在。

等轴测图设定平面图的两根轴线与水平基线成30°夹角，并且把高度轴线绘制在平面图轴线上。这就意味着被表示的物体不会发生前述三种方法中的混乱情况，但是等轴测图必须精心绘制。

对正二轴测图而言，平面图的两个轴线与水平基线的夹角设定为7°和42°。其中，夹角为42°的轴线长度应按0.5或者0.7的比例缩短。

透视图

透视图不同于轴测图、等轴测图和正二轴测图，因为在后三种投影图中，同一轴线方向上的线并不是表现为平行，这与透视图中的情况刚好相反。由于透视图一般不用在建筑设计中，而仅仅用于表现设计意图，且透视图属于投影几何学领域，因此本书中不作深入研究。

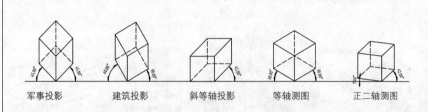

图 6：
轴测图、等轴测图和正二轴测图

P14　**表现原则**

P14　　　　　**辅助工具**

基本上，施工图的绘制有两种不同的方法：
— 手工绘图；
— CAD。

绘图桌　　　　手工绘图是在配有一副滑标尺的特殊绘图桌上进行的，这种滑标尺被设置成合适的、可以调整的角度。绘图桌也可以配有绘图轨道，这种轨道可以在桌面上旋转，或者沿着张紧的钢丝垂直滑动。使用这两种不同的方法均可以画出平行线或者成直角的线。

绘图水笔和
铅笔

　　　　手工绘图通常是用铅笔或者墨水钢笔来完成的。铅笔有各种不同的硬度级别，可以画出不同宽度和视觉效果的线条：铅笔越硬，线越细。因为铅笔越硬，笔头在图纸上的磨损越少。所以，绘图需要不同硬度级别的铅笔，以绘制不同宽度的线条。

　　　　墨水钢笔有多种形式（比如有无笔芯）及多种笔尖宽度。此处注

注释：
　　施工图绘制中用到的铅笔等级从 B（软）、F（中等硬）、H 到 3H（各种不同的硬度级别）。绘图时应首先使用硬度大的铅笔，以避免涂污了那些较软的、浓度较大的线。

注释：
　　各种颜色的墨水钢笔分为有笔芯和无笔芯两种。在绘草图时，无笔芯的墨水钢笔更便宜；但如果需要绘制大量的图纸时，无笔芯的墨水钢笔会更贵一些。画出的线越细，所使用的钢笔越容易在长时间放置后发干堵塞。有时，笔尖的色素在浸水之后能够被洇湿。如果需要擦除已经画好的线，可以使用特殊的墨水消除剂。采用刀片刮除图形时，应做到动作迅速。

图7：
标准辅助工具

直尺 和三角尺	释中所提的笔尖可以单独购买。 使用多种直尺、量角器、三角板、三角尺以及曲线板可以使得绘图变得更容易。直尺、三角板以及可调三角尺用于按几何尺寸的绘图。平面图上长度的测量通常采用三角尺和具有六种不同尺寸比例的三角尺来完成。
绘图模板	几乎所有的标准绘图符号都可以在绘图模板上找到（如家具、电器连接或浴室设施）。也有适用于铅版印刷的绘图模板。对于不同的线宽和比例尺，有相应的绘图模板。
CAD 程序 💡	CAD 绘图利用计算机完成。绘图者需要一个专门用于施工图绘制的 CAD（计算机辅助设计）程序。市场上有各种各样的程序可供使用，但在简易性、性能和价格等方面有所差别。几乎所有的供应商都提供针对学生和学校使用的版本。
P16 图纸规格	**图纸规格和类型** DIN 476 – 1 或者 ISO 216 中以 $1:\sqrt{2}$ 的页面比例为基础，定义了不同的图纸规格。采用这种页面比例的优点在于：大号图纸在分成小号

> 💡
> **重点：**
> 在用户使用程序工作之前，最好考虑一下各供应商所提供的程序。即使都是学生版本，价格的差别也还是相当大的。另外，在学生或者同事之间使用相同的程序也是很重要的，因为这样有助于彼此交流工作经验。

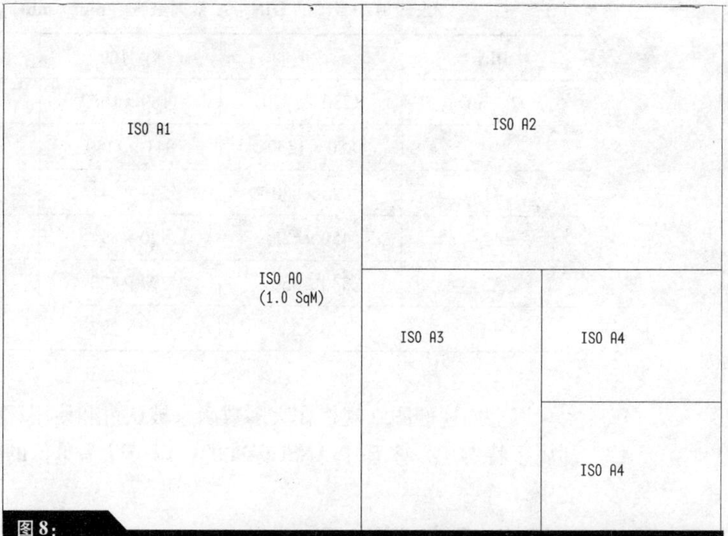

图8：
ISO – A 系列中的图纸规格关系

图纸时，不会有丝毫浪费。

在 DIN 或者 ISO 标准范围内，有各种系列的图纸规格。其中，广泛用于设计图的是 DIN – A 或 ISO – A 系列。

由于图纸大小在按尺寸裁剪及折叠中会有所损失，因此在修剪和未修剪的图纸规格之间是有区别的。在纸张市场上也有诸如"加大 DIN A3"类别的图纸，但是这些类别的图纸都是由印刷厂生产的，并没有进一步标准化。

ISO/DIN A – E 系列（mm×mm） 表1

	A –	B –	C –	D –	E –
2 – 0	1189 × 1682	1414 × 2000			
0	841 × 1189	1000 × 1414	917 × 1297	771 × 1091	800 × 1120
– 1	594 × 841	707 × 1000	648 × 917	545 × 771	560 × 800
– 2	420 × 594	500 × 707	458 × 648	385 × 545	400 × 560
– 3	297 × 420	353 × 500	324 × 458	272 × 385	280 × 400
– 4	210 × 297	250 × 353	229 × 324	192 × 272	200 × 280

未裁剪和裁剪的 DIN – A 系列纸张（mm × mm） 表2

DIN	未裁剪的	裁剪的	边距
2 – A0	1230 × 1720	1189 × 1682	10
A0	880 × 1230	841 × 1189	10
A1	625 × 880	594 × 841	10
A2	450 × 625	420 × 594	10
A3	330 × 450	297 × 420	5
A4	240 × 330	210 × 297	5

以上描述的图纸规格在大多数国家被认可和使用；但是北美地区却属于特例，一些基于 ANSI 标准的，以英寸为单位的图纸也在使用之中。

ANSI 图纸规格 表3

系列	工程师的图纸	建筑师的图纸	工程师的图纸	建筑师的图纸
A	8½ × 11	9 × 12	216 × 279	229 × 305
B	11 × 17	12 × 18	279 × 432	305 × 457
C	17 × 22	18 × 24	432 × 559	457 × 610
D	22 × 34	24 × 36	559 × 864	610 × 914
E	34 × 44	36 × 48	864 × 1118	914 × 1219
F	44 × 68		1118 × 1727	
	in × in		mm × mm	

图纸类型

与图纸规格一样，图纸也分成不同的类型。通常，描图纸用于手工绘图，因为描图纸的优点在于可以将其他需要描绘的图纸放在下面。这使得施工图绘制大为简化（如绘制上部楼层图或剖面图）。描图纸还可以通过使用蓝图来简化原图的复制。

在图纸保存方面，常使用拉延润滑膜，因为这种拉延润滑膜即使在较高的温度时仍可以保持其形状。因此，即使在很长一段时间后，仍然可以从图上得到可靠的量测数据。

在用 CAD 程序绘制工程图纸时，常用的绘图纸是卷轴白图纸或片材白图纸。因为铜版纸、照片纸或光面纸具有高品质的表层,因此常用于影像图。

P18

比例

在第一章(工程图纸的类型)中提到的每一种图纸类型都是实际建筑物按一定的比例缩小得到的,即按一种特定的比例来绘制。在每张图纸上必须标明比例尺,以"比例"一词和两个用冒号隔开的数字的形式来表示(例如"比例1:10")。

比例的定义

比例描述的是图纸上一个绘图元素的尺寸和它实际尺寸之间的关系。以下三种比例应当加以区分:

—— 原始比例(比例1:1),即自然比例;

—— 放大比例(比例x:1),在此比例下,按一定的倍数绘制比实际尺寸大的图形元素;

—— 缩小比例(比例1:x),在此比例下,按一定的倍数绘制比实际尺寸小的图形元素。

如此一来,按比例1:100绘图得到的墙体尺寸就相当于其原有尺寸的1/100。

标准比例

由于实际建筑物比图纸要大很多,因此施工图中总是采用缩小比例。随着设计过程中精度和细节的增加,绘图比例的缩小程度有所减少,因而图纸中所描述的目标物也会变得更大。

位置图和粗测绘图通常都是以1:500(或者1:1000)的比例来绘制,设计图是以1:200或者1:100的比例尺来绘制的。对于施工图,通常使用的比例尺有:1:50、1:25、1:20、1:10、1:5、1:2以及1:1(见"制图步骤"一章)。

比例的转换

如果一堵5.5m长的墙以1:50的比例描述出来,其长度必须除以这个缩小因子:即5.5m/50 = 0.11m,绘图时的长度也就是11cm。当一个已经以某一缩小比例绘制好的物体必须转换成另一个不同的比

注释:

为了把原始长度转换成标准施工图绘图比例,最好使用三角比例尺或者按照下面的方法来计算比例长度:

—— 1:10、1:100、1:1000 的比例尺——通过数字零移动小数点,对于1:100的比例把比例尺单位从m变到cm,对于1:1000的比例把比例尺单位从m变到mm;

—— 1:5、1:50、1:500 的比例尺——仿照上面移动小数点并把得到的数字乘以2;

—— 1:20、1:200 的比例尺——仿照上面移动小数点并把得到的数字除以2。

	例时,绘图就会变得很困难。如果一扇门以 1:20 的比例绘制在图上的长度是5cm,当需要以 1:50 的比例来表示这扇门时,就必须进行两种比例的转换,即长度就变成 5cm × 20/50 = 5cm/2.5 = 2cm(其中2.5为转换系数)。
CAD 程序中的比例	CAD 程序简化了比例换算问题。在 CAD 绘图中,建筑物通常都是以 1:1 的比例来进行绘制,即一堵 5.5m 的墙就绘制成此长度。图纸则是通过输出比例或者参考比例而另外得到的,即确定随后图纸打印和输出时的比例。绘图的线宽和文字大小在显示器上显示时,也是采用了这种参考比例;如此,最终的结果一目了然。
P19	线
	工程图纸由很多的线组成,根据线的类型和宽度可以区分所表示的物体。虽然线型和线宽的意义随着比例的不同而改变,但仍有必要对线型和线宽进行区分。
线型	有四种基本类型的线:连续线、短划线、点划线以及点线。从这些基本线型可以引申出介于它们之间的其他线型。
线宽	虽然通常只有宽度超过 0.7mm 的线才被使用,但下述线宽都是惯用的:0.13mm、0.18mm、0.25mm、0.35mm、0.5mm、0.7mm、1mm、1.4mm、2mm。
连续线的使用	所有的可见物体以及剖面图的可见边界都使用连续线;剖面区域的分界线也是用连续线识别的。当建筑物的部分被剖切出来,以 1:200 和 1:100 的比例显示在剖面图上时,通常使用 0.25~0.5mm 宽的连续线;而当比例大于 1:50 时,建议使用 0.7~1mm 宽的连续线。辅助性建筑、尺寸线、次要的俯视图或者平面图中使用更细的连续线:比例为 1:200 或 1:100 时,线宽为 0.18~0.25mm;比例大于 1:50时,线宽为 0.25~0.5mm。
短划线及点线的使用	短划线用于被遮挡的建筑物边缘(如楼梯详图中踏步以下部分)。当比例是 1:200 和 1:100 时,线宽通常为 0.25~0.35mm;当比例大于 1:50 时,线宽为 0.5~0.7mm。
	轴线以及剖面图的取向线常使用点划线。由于剖面图取向线要求易于辨认,因此当比例为 1:200 和 1:100 时,线宽为 0.5mm;当比例大于 1:50 时,线宽为 1mm。另一方面,对于轴线而言,当比例为 1:200 和 1:100 时,线宽为 0.18~0.25mm;当比例大于 1:50 时,线宽为 0.35~0.5mm。

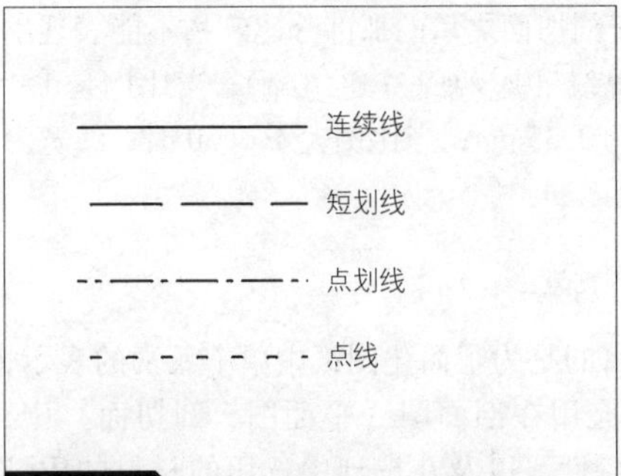

———————— 连续线

— — — — 短划线

·—·—·— 点划线

· · · · · 点线

图 9:
线型

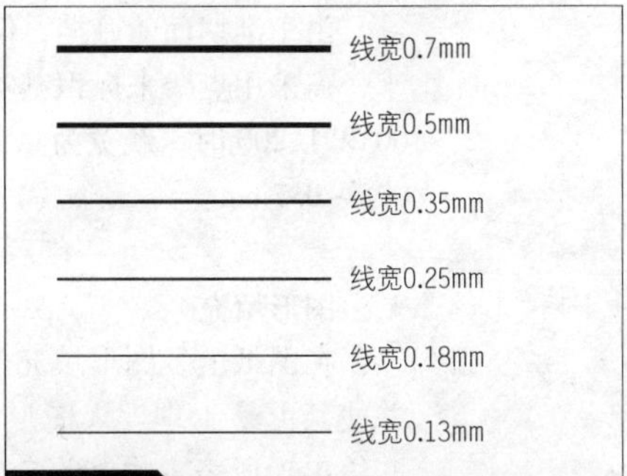

———————— 线宽0.7mm

———————— 线宽0.5mm

———————— 线宽0.35mm

———————— 线宽0.25mm

———————— 线宽0.18mm

———————— 线宽0.13mm

图 10:
线宽

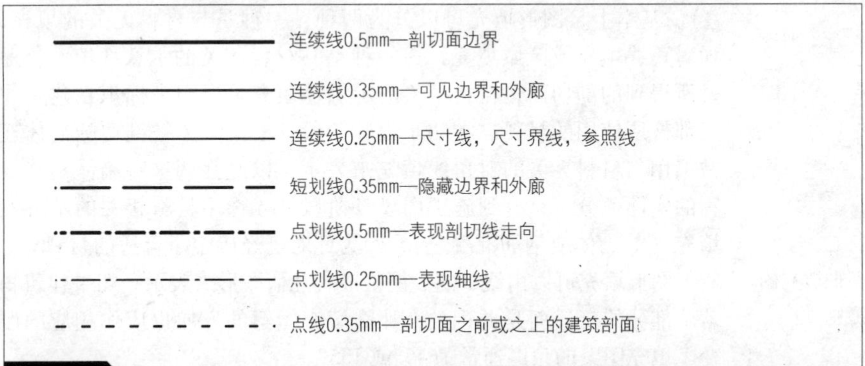

———————— 连续线0.5mm——剖切面边界

———————— 连续线0.35mm——可见边界和外廓

———————— 连续线0.25mm——尺寸线，尺寸界线，参照线

— — — — 短划线0.35mm——隐藏边界和外廓

·—·—·— 点划线0.5mm——表现剖切线走向

·—·—·— 点划线0.25mm——表现轴线

· · · · · 点线0.35mm——剖切面之前或之上的建筑剖面

图 11:
线型和线宽，比例尺：1:100

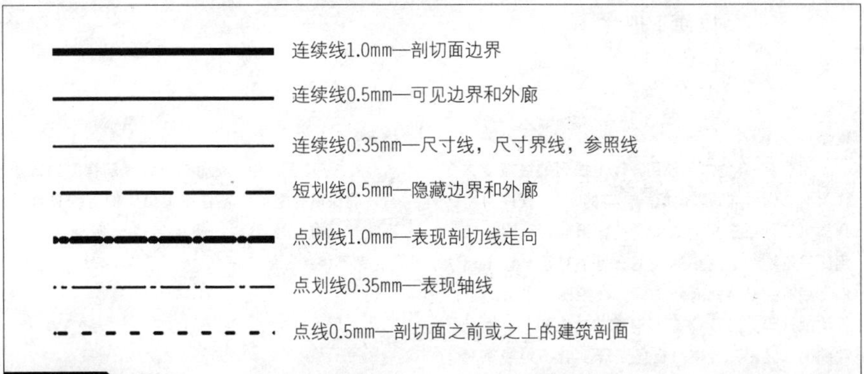

———————— 连续线1.0mm——剖切面边界

———————— 连续线0.5mm——可见边界和外廓

———————— 连续线0.35mm——尺寸线，尺寸界线，参照线

— — — — 短划线0.5mm——隐藏边界和外廓

·—·—·— 点划线1.0mm——表现剖切线走向

·—·—·— 点划线0.35mm——表现轴线

· · · · · 点线0.5mm——剖切面之前或之上的建筑剖面

图 12:
线型和线宽，比例尺：1:50

由于剖切面的遮挡，位于剖切面之后的那部分建筑物不能表现出来，需采用点线来标识其边界线（见"投影分类"一章）。当比例是1∶100和1∶200时，线宽为0.25~0.35mm；当比例大于1∶50时，线宽为0.5~0.7mm。

图形填充

在图纸上，图形填充的目的是为了简化图纸中单个元素的表示，并使其更易于理解。图形填充用在剖面图（平面图、剖切面）中，并给出了所表示物体的自然特性，以及在设计中选用的材料的质量和组分等信息。当从建筑物上得到剖切面时，被线围绕封闭的区域将被填充。关于图形填充的方法已经被汇集成德国国家标准（见附录）。基本上，图形填充可以分为两种，一种是与材料无关的填充，如对角线填充或区域填充；另一种是与材料相关的示意性填充。通过所得到的剖切面，材料相关的示意性填充可以用来标识在建筑物某部位所使用的材料。在初步设计阶段，平面图上被剖切的墙体通常采用与材料无关的对角线填充来表示，以此强调这些墙体是建筑物的实体部分。只有到施工图设计阶段，才选用材料相关的示意性填充（如砌体或钢筋混凝土），因为此时已经确定了合适的材料。

图形填充的原则

图形填充可以由线、点、网格或者几何图形来表示。如果建筑物若干部分的交界面是并列的，那么填充图案的方向也应作相应的改变。填充图案的角度通常为45°或135°。

表示与材料无关的建筑部分交界面的最基本的填充图案是倾角为45°的连续线填充。狭窄的剖切面应被涂黑，如钢结构中的横截面，以便于理解。

提示：

由于CAD的广泛使用，以上提到的线宽应该作为一个指导准则来理解。当前的CAD程序为用户提供了可自定义的画笔种类，因此，有时会使用比手工绘图中更小的线宽。为了对某一输出比例下线宽的影响能有所估计，在绘图之前务必进行测试打印。因为CAD具有缩放功能和与比例无关的图形显示，会使得在显示屏上看不到线宽对打印结果的影响。

提示：

在CAD程序中，图形填充按比例缩放的方法是：确保填充图案能在任何输出比例下均具有可视效果。在这本书中，使用了"比例相关"和"比例不相关"的术语。

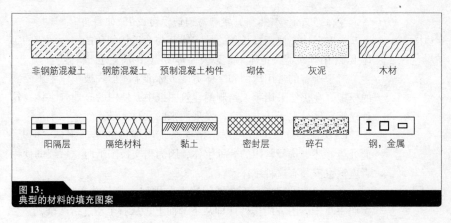

图 13：
典型的材料的填充图案

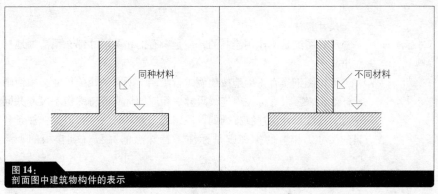

图 14：
剖面图中建筑物构件的表示

P22
文字标注

如同线、面和图形填充一样，对于绘制一幅完整的图纸，适当的文字标注也是必不可少的。文字标注的详细程度是根据与绘图比例的关系来选择的；此处，文字标注有助于工程制图（如说明尺寸、房间数目、材料信息等）。

选用的字体必须明确、易于辨认，所以通常选用一种标准字体。标准字体（也称为 ISO 字体）是国际上使用的大写及小写文字标注的字体名称。根据字体尺寸和倾斜角度的不同，标准字体被分成四种不同的形式。按字体宽度可分为两种：

字体 A - 狭窄，线宽是文字高度的 1/14；

字体 B - 中等宽度，线宽是文字高度的 1/10。

按字体倾斜角度可分为两种：

字体倾角 v - 正体字，字母垂直于阅读方向布置；

字体倾角 i – 斜体字，字母与阅读方向成 75°倾斜角。

字体样式的使用

标准字体中使用最长久的是字体 B 与合适倾斜角的结合而成的字体 Bv（中等线宽，正体字）以及字体 Bi（中等线宽，斜体字）。

对于所有的常用比例和字体宽度，标准字体形式被用作手工绘图的模板。在仍然采用手工绘制的建筑图纸中，使用的是建筑字体。这种建筑字体只限于从方形字体发展而来的大写字母。

CAD 程序通常可以使用由操作系统所提供的全部字体。但是，此处最好还是选择常用的一种字体，因为当交换图纸时，后续的使用者应该也安装有这种字体。

文字始终按照平行于图纸阅读方向排列；或者在逆时针旋转后，按照垂直于读图方向排列（即从下部而上、从右至左进行阅读）。

P24 尺寸标注的原则

尺寸标注

尽管图纸是按比例绘制的，但是所有的相关尺寸都必须清晰地标注出来。

无论图纸是否正确地按比例进行绘制，所有的相关尺寸必须清晰地标注出来。尺寸标注由连续尺寸、相应的标高或特殊指示尺寸共同完成。连续尺寸由连续排列的区段组成，并且每一个区段都标注了各自的尺寸。相应的标高则是标注了特殊点的高程（如楼板的上表面）。

在德国国家标准中列出了在图纸上进行尺寸标注的指导准则。

```
ABCDEFGHIJKLMNOPRSTUVWXYZ
abcdefghijklmnoprstuvwxyz
1234567890 10 [ ( ! ? : ; - = ) ]
```
正体字

```
ABCDEFGHIJKLMNOPRSTUVWXYZ
abcdefghijklmnoprstuvwxyz
1234567890 10 [ ( ! ? : ; - = ) ]
```
斜体字

图 15：
标准字体 Bv 以及 Bi

（见附录）

尺寸链

尺寸链的结构　　尺寸链由以下部分组成：
- 尺寸线；
- 尺寸界线；
- 尺寸起止符号；
- 尺寸数字。

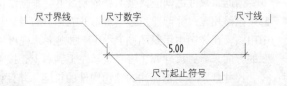

图 16：
一个尺寸链的元素

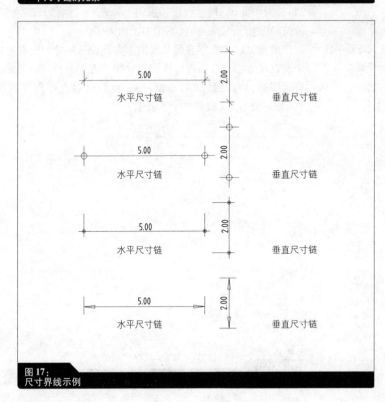

图 17：
尺寸界线示例

尺寸起止符号	尺寸线、尺寸起止符号及尺寸界线均采用连续线表示。尺寸线平行于被标注的结构部分。尺寸界线垂直于尺寸线，用以标明被标注的轴线、边界或者线。

尺寸起止符号在尺寸线上定义了被标注长度的外部点。即使尺寸界线在原则上定义了这些点，然而对于图纸上的交叉线来说，无论墙体剖切面上的粗线是否包含了尺寸起止符号，这种定义仍不清晰。

因此，尺寸之间应该被明确地划分开。根据绘图比例，斜线和圆都可以被用作尺寸起止符号（如起止符号在设计图中用直线、施工图中用圆，在详图中用封闭圆），但是原则上来说可以自由选择。从读图的方向来看，尺寸起止符号是从左下到右上，按45°角来绘制的。 |
| 尺寸中的尺寸数字 | 尺寸数字等于被标注的结构构件的长度，并且相应地也可以确定尺寸起止符号之间的距离。当距离大于1m时，尺寸数字的单位是 m（例如1），当距离小于1m时，尺寸数字的单位是 cm（例如99 或者25）。对于 mm，尺寸数字是采用上标表示（例如1.25^5 或36^5）。而单位符号，如 m 或 cm，就不用表示出来了。 |
| 尺寸数字的位置 | 通常，尺寸数字在尺寸线的上方，位于尺寸起止符号的中央。对于洞口尺寸，洞口高度标注在尺寸线下方（见图19）。如果洞口位置还有附属的窗台（例如对于窗户），窗台的高度则直接在洞口内给出（例如用符号 SH 或者 $S = 75$ 表示）。 |

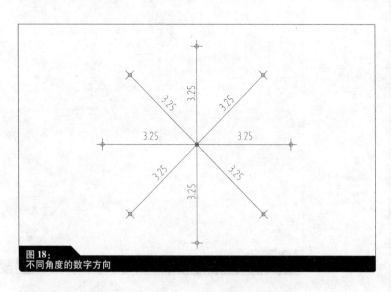

图18：
不同角度的数字方向

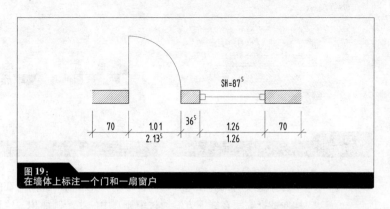

图19：
在墙体上标注一个门和一扇窗户

在尺寸线内的尺寸标注是不常用的：这种标注是前述标注的一种替代方式。在这种标注方式中，尺寸数字直接写在尺寸线上面，从而省略了部分尺寸线。

当尺寸起止符号之间的距离比较小时，会出现一些问题。也就是只有很小的空间来标注尺寸数字（例如轻质墙、墙面的设备）。此时，尺寸数字可以直接标注在尺寸起止符号的旁边（见图25～图27）。尺寸线和尺寸数字一定不能重叠，以此保证尺寸的易读性。

标高

立面图/剖面图上的标高

标高包含了楼层、窗台、净空高度以及相对于平面 ±0.00 的高度。标高通常标注在已完成楼面结构上表面的入口区域。测量员经过测量，将这些点与平均海平面标高或者零标高平面联系起来，由此建筑物就可以做到准确的高度定位。因此，在每张图纸中，都应当规定参考零平面和 ±0.00。所有的标高符号与起始点 ±0.00 的关系用符号 "+" 或者 "-" 表示。

绘图时，平面图及俯视图上的标高与剖面图及立面图有所不同。

平面图上的标高

立面图和剖面图上的标高采用标高符号表示：标高符号是一个等边三角形，标高符号和标高数字直接标注在结构构件中，或者标注在一个辅助的尺寸起止符号上（例如标注在建筑物的外部）。

对于结构面的标高，采用黑色的实心三角形；对于完成面的标高，则采用仅有轮廓线的三角形。这样，即使在很小的图纸上，标高也可以清晰准确地标注出来。

平面图和俯视图上的标高也用三角形标志来表示（结构面用实心三角形，完成面用用空心三角形）。但是，最常见的标高符号是一个

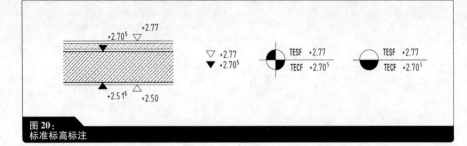

图 20：
标准标高标注

带有穿透线的圆。在这种标高符号中，完成面的标高写在线上方，结构面的标高写在线下方。通过填充其中的半个圆面积，这种标高符号可以直接辨认出来。另一类标高符号的使用文字标注 TECF（完成楼面上表面）和 TESF（结构楼面上表面）来表示。

角度和弧形尺寸

角度尺寸　　如果结构构件互相之间不是成直角，那么相关的角度就必须标注出来。角度标注通常表示为：标出的角度数和符号∢。但是，角度尺寸也可以采用两端带箭头、角度数居中的圆弧来标注。

弧形尺寸　　对于有圆角的结构构件，应当给出圆弧的尺寸，例如定义弧形钢筋混凝土墙体的展开长度。在其他的一些构件中，弧形尺寸对于确定和计算尺寸也是必不可少的（如墙体的连续尺寸、踢脚板长度等）。弧形尺寸包括了平行于实际弧形结构的圆弧线（如尺寸线是弧形墙体的同心圆弧线）。前述的尺寸起止符号可以用于弧形尺寸标注，端部的箭头也可以作为其尺寸起止符号。

特殊尺寸符号

如果需要标注特殊尺寸，通常是将尺寸直接写在特殊结构构件上面。特殊尺寸可以用缩写（例如 SH 表示窗台高度），或者符号（例如 Φ 表示直径、□表示一个矩形断面）来标识。用大写字母 R 加上尺寸数字可以表示半径大小，用大写字母 M 来表示螺栓和拉杆。为了简化表示，高度和宽度也可以用缩写来表示（例如 W/H 12/16 表示一个木梁的宽、高分别是 12cm 和 16cm）。

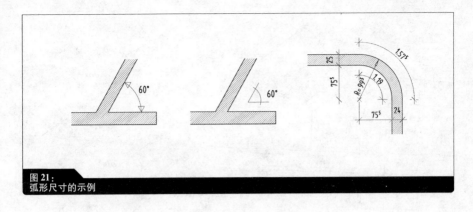

图 21:
弧形尺寸的示例

P30

制图步骤

建筑图纸可分为两类,第一类图纸表现的是设计思路的形成阶段:从规划到方案许可。第二类图纸表现的是建筑物的施工阶段,包含与建筑物相关的图纸。因此,初步设计、正式设计、特殊部位的详图和施工图之间有很大差别。

在设计文件中,包含了针对相应图纸要求的部分。这些文件可作为客户或当地的权威建设部门的决策依据;专业的设计者可以将这些资料作为自己的设计依据。这些文件中包含混凝土建筑施工说明,用来指导专业的单位进行施工。图纸内容的范围和精度取决于制图目的、特点和比例。图纸包含范围越小,图纸上结构构件就越大,因此标注的尺寸就越详细。

P30

土地利用图

定项基础

城市和市镇的总体规划图(通常比例尺为 1:1000)可以提供土地利用图,以此作为设计的概览和基础。这张图上,尺度标注上会有一些轻微误差。

建成建筑图

如果打算对一栋已有建筑进行改造,那么必须调查清楚其现状,并将其绘制成图,以作为改造工作的基础。随后,一系列的设计图纸都将以此作为依据。竣工图的精度和准确度很大程度上依赖于建筑物的长远考虑。如果需要已建房屋进行小规模的扩建,且没有细节上的特殊要求,那么只要记录待扩建部位的长、宽、高就足够了。但是,如果需要考虑古迹保护措施,那么图上的尺寸标注就必须详细,并补

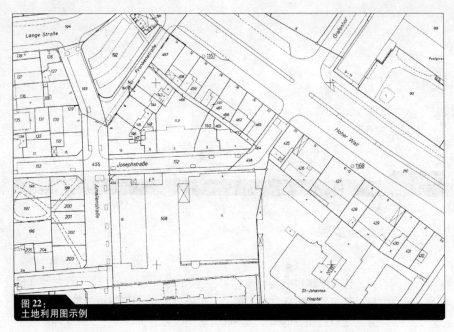

图 22：
土地利用图示例

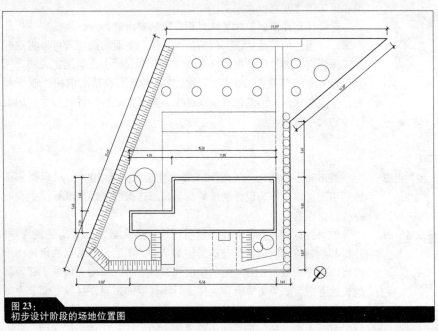

图 23：
初步设计阶段的场地位置图

充关于建筑物表面和风格特点等内容。

初步设计图

初步设计图可以全面地表达设计者的思路。初步设计图分为两种：一种是用来表示未来设计的建筑图，另一种是用于给客户解释设计思路的概念图。初步设计图的目的是明确和解释藏于设计之中的思路。因此，初步设计图能表达作者的设计方法，并且在表现手法上也有很大的自由度。另一方面，有关权威部门也可以参考初步设计图，以明确了解建筑物。因此，初步设计主要也表达了建设部门的基本思路。

初步设计图只给出建筑物最基本的信息，如建筑物的形状和尺寸。通常情况下，初步设计图的比例为1:200；对于大型工程项目，图纸比例为1:500；而场地位置图的比例则更小（比例为1:500或1:1000）。

场地位置图表示了建筑物的地理位置以及周围环境。场地位置图用于确定建筑物的位置，并由此得名。该图大致介绍了建筑物规模、朝向、地形特征及用途，还包括了必要的相邻地块等。

绘制深入设计的施工图时，明智之举是首先着手于那些与场地位置图相匹配的地面层平面图。如果首先完成地面层平面图，那么它就能够为上部楼层的平面图提供最好的绘制方法（在绘图板上或CAD程序中均能如此）。

完成平面图后，剖面图就相对简单多了。首先，在平面图上画出选择的剖面线；然后，将剖面向变成水平方向，从而得到剖面图的基线。这条基线是此剖面图上的标高零线，即建筑物入口处的高程。剖面图上所有垂直向上（上部楼层）或垂直向下的标高（地下室）都以这条基线为参考来确定：绘制墙体的剖面时，将平面图上的剖面线与墙体的相交线在垂直方向上进行延伸，延伸到达合适的标高位置即可。立面图应以剖面图上的标高为基础，把平面图放在下面，以此简单画出建筑物的外轮廓线，然后加上窗户、门洞和地面等即可。（见图24）

初步设计的目的是明确和表示建筑物的容积、空间分布、与周围环境的协调等，这些内容是通过临时配置和临时尺寸来表示的。由于结构构件在表示时通常体现不出其材料的特性，因此，首先只能确定哪些是相交构件。

在立面图上，所有可见边界都是通过连续线表示，连续线粗细取决

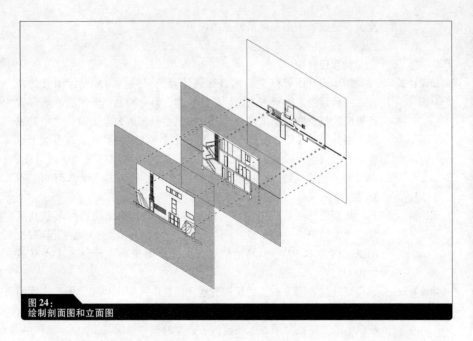

图24：
绘制剖面图和立面图

于绘图比例、相关性（例如，与门上的把手相比，墙体更重要）和在建筑物中的详细程度。因此，外墙和墙体的洞口轮廓线被强调得更多。

为了说明设计效果，需要在俯视图和立面图上完整地画出厕所、厨房和家具等详图。家居设备详图对于缺乏经验的用户而言很重要，尤其房屋大小的尺寸比例，这些家居设备详图可以让顾客更好地了解房屋的比例和大小。

设定比例　　图上的树、人和外部区域情景等为建筑物提供了一个显而易见的背景，这明确了建筑物是如何与周围环境相协调的。不管读者是否了解这方面的内容，图中的比例和尺度可以让他们很好地明白建筑物的尺寸和比例。这些就是所谓的设定比例，在个别项目上也称为比例设定目标。这些额外添加的部分主要用在初步设计和正式设计阶段，用于项目竞标和效果图上。

尺寸标注　　在初步设计图纸上，仅限于粗略的尺寸标注。在平面图上，给出了外部尺寸和重要房间的尺寸，以便更好了解建筑物的总体大小和房间尺寸。个别的凸出和凹进部分，门、窗洞口的尺寸通常不会标注出来。

在立面图上，仅有那些重要的部位（如屋檐、屋脊）才给出标高；在剖面图上，仅给出房屋和楼层的高度。

文字标注　　文字标注也仅局限于简单表示房屋的功能和估算的平方米面积。

提示：

图中给出的是 Vaucresson 地区早期的住宅建筑，由勒·柯布西耶设计（1922 年）。作者从勒·柯布西耶的图纸上拷贝下来该图，但是为了此处的相关点，作者改进了一些标注和细节，完善了尺寸标注。其中，部分门、厕所等尺寸已不再适应于当前规范，因此不能作为读者绘图的模板。

注释：

在学生制图和竞赛中提交作品时，采取以下手段会有所帮助：画上标题栏并列出设计思路的关键点，加以图形说明。

P39
给客户的
表现图

表现图

为了表达设计内容，表现图的准备是独立于传统的施工图纸的。表现图通常是在初步设计完成后才开始进行，以便能确定深入设计的意图：因为表现图目的在于说服那些具有设计思路和概念的特殊群体，因此效果图在绘制中应把握这一要求。

例如，如果一个没有经验的客户，他无法将二维绘图联系到三维空间上来，那么表现图就可以通过三维图或透视图来加强客户的空间感。

有时，客户必须向第三方展示设计内容，需要利用一些栩栩如生的材料来说服他们。此时需要三维表现图，透视图，区域情况、交通情况和影响范围的图解表现图，及以平面图和剖面图为基础的类似材料等。

建筑竞标图

如果设计者或学生参与了方案竞标，那么他们的设计思想必须以某种方式表现出来；采取这种表现方式，将给评标人员带来视觉震撼。通常，评标人员由专家和非专业人士组成，因此双方的要求都应考虑周全。一般情况下，对于每一个提交的竞标方案评标的时间很少，因此每一个评委都必须尽快了解清楚。因此，使自己的方案能够脱颖而出也是非常重要的方面。

学生表现图

对于学生的设计作品，负责的大学老师应能够理解学生的设计思想。表现图应使有资格的专家信服，这些专家具备从表现图中抽象和想像的能力。因此，提交作品的学生更愿意选择概念性的方法表达他们的设计思想，而不愿选择那些面向客户时的表达方式。例如，图解表现图上可能略掉了比例设定特征，而这些比例设定特征对于非专业人士是必不可少的。

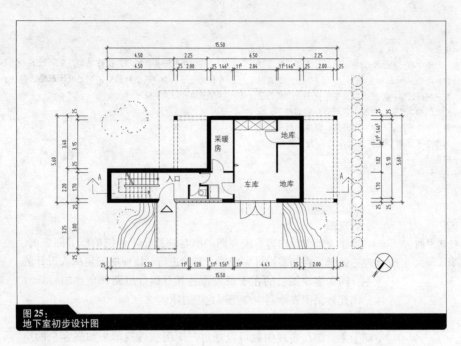

图 25：
地下室初步设计图

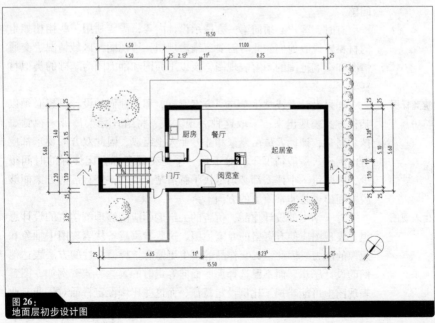

图 26：
地面层初步设计图

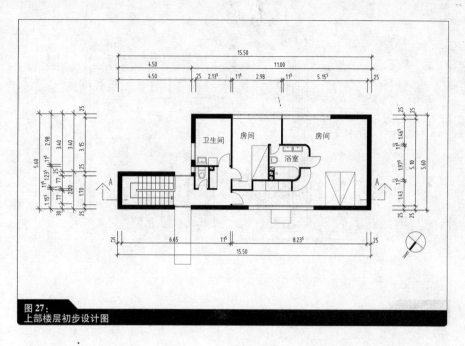

图27:
上部楼层初步设计图

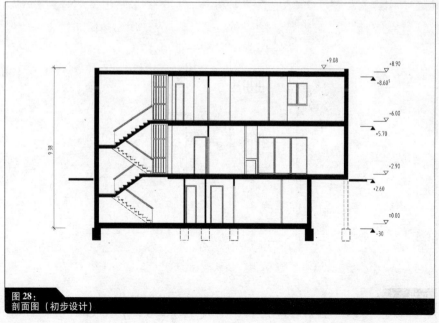

图28:
剖面图(初步设计)

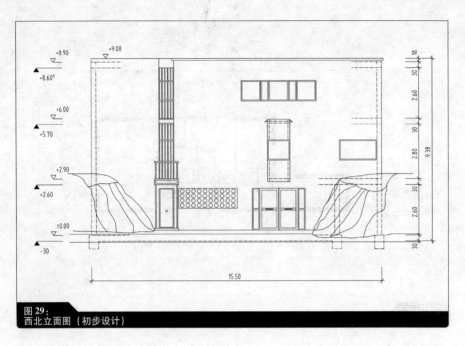

图29:
西北立面图(初步设计)

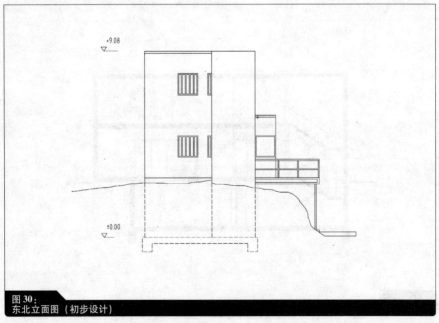

图30:
东北立面图(初步设计)

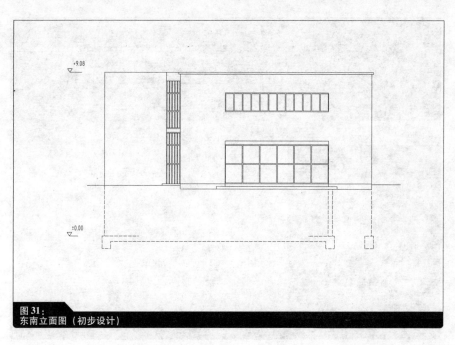

图 31：
东南立面图（初步设计）

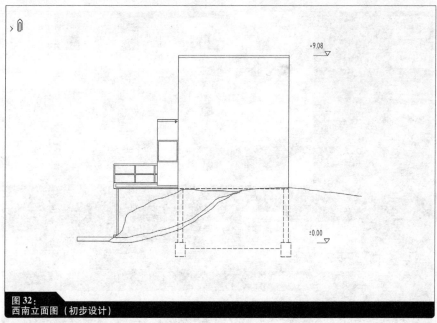

图 32：
西南立面图（初步设计）

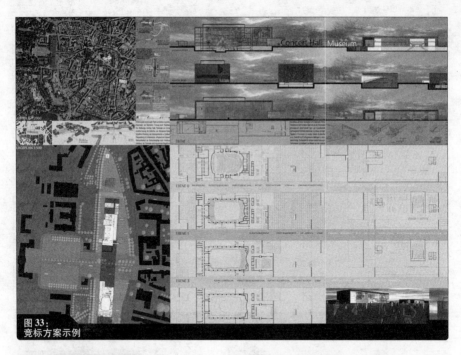

图 33：
竞标方案示例

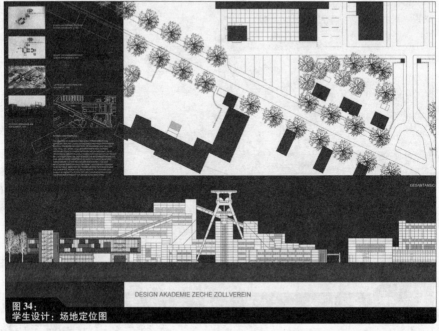

图 34：
学生设计：场地定位图

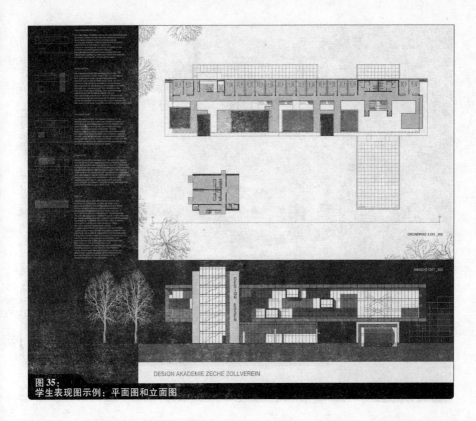

图35：
学生表现图示例：平面图和立面图

表现图的内容	除了包含传统的场地定位图、平面图、剖面图、立面图等相同以外，表现图还包括建筑物内部和外部的三维视图。图解方法表示的结构构件也有助于表达设计思路和功能。
比例固定的表现图	提交的图纸通常采用固定的绘图比例。对于表现图，比工程图纸具有更大的自由度。在固定比例表现图中，尺寸可以缩为最小值，图形表示的结构构件和整个图纸区域都可以从根本上缩小。这种表示方法，不会限制设计者的创造力。对于考虑此项工程的设计者而言，清晰、简便地记住有关内容以及给出合适的表现图显得更为重要。
P42 设计图纸的目的	**设计图纸** 设计制图是初步设计的深入发展阶段，对于获得方案许可的建筑，建筑师和客户最终确定设计的形状和尺寸。所以，设计图纸上应该体现出来图纸授权者所考虑的相关重要内容。这个阶段也包括其他图纸，诸如结构工程图和公共设备图。这就意味着基本结构单元（如

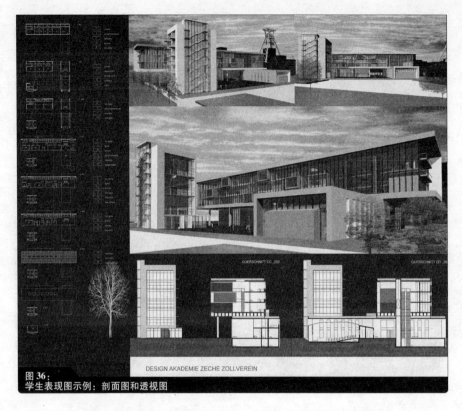

图36：
学生表现图示例：剖面图和透视图

承重墙）必须画出来。

绘图比例　　住宅的设计平面图房屋比例采用1:100；必要时，大型建筑的平面图比例采用1:200或者1:500。例如一个大型的工业厂房以比例1:100画出来，就需要大量的A0号图纸，平面图也不能作为一个整体画在一张图纸上。但是，如前所说，由于设计图纸的目的是提供易懂有用的图纸，以作为深入讨论的根本。所以选择一个小精度、大比例表现方法非常有意义。

墙体的表示　　在设计图纸阶段，为了定义墙体的材料，采用了与材料相关的图形填充表示墙体，如钢筋混凝土墙、砌体墙或清水墙。墙体线的粗细区分出承重墙和非承重墙。此阶段中，通常不会显示墙体的表面情况（例如内部抹灰地面层）。在图纸上，通过标明洞口尺寸和窗台，对门、窗等进行准确绘制。在设计图纸上，应标明门的开启方向，由此表示建筑物内部的流线。

征询结构工程师的意见后，可以绘出基础的建筑详图（独立基

楼面和顶棚的表示	础、冻土墙、条形基础），包括正确的深度和宽度。如果有必要说明截面，则看不到的部分可以用虚线表示。 在剖面图上，为了确定结构的标高，楼面和顶棚分为结构部分和完成面，其中结构面由填充图案来表示所用的材料，完成面由结构面上部的边界线来表示。
不可见部分的表示	由于平面图的水平剖面位于楼地面上部1.5m处，因此在设计图中看不到剖面以上的结构。但是，为了更好理解空间的几何特性，这些不可见部分也应该表示出来。这些不可见部分包括了大大小小的梁，这些梁将空间分成了几部分（在平面图上，梁的尺寸应该直接表示出来，如B43/35）；还包括楼梯，应该表示出楼梯的上部，以更好理解踏步形状。同样，还包括休息平台上和不可见的楼梯段。在立面图上，内部的承重墙和楼层用虚线表示。
立面图上的附加信息	立面图上应该表示以下部分：窗户区域和打开方式（建筑法规中未规定），内部的梁，窗口高度，阳台，窗台，凸出及隐蔽处，屋面形式。
场地示意图	在建筑物修建前后，已有的场地示意图应尽可能精确，因为它关系到建筑监督局和土方工程的出入口，建筑标高体系也从此图上来确定。

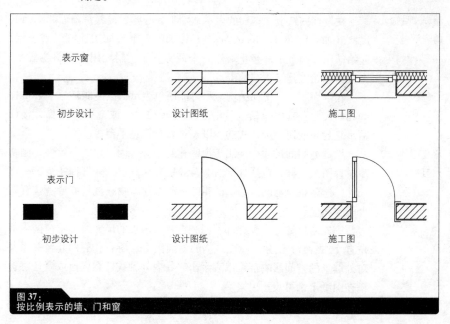

图37：
按比例表示的墙、门和窗

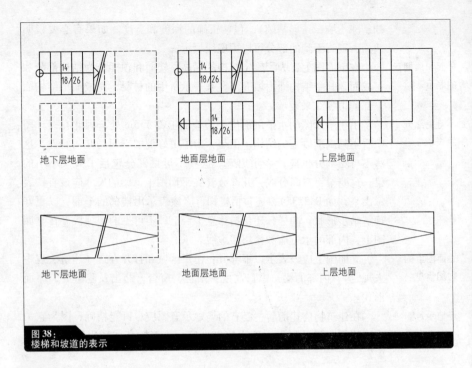

图38:
楼梯和坡道的表示

地下层地面　　　　地面层地面　　　　上层地面

地下层地面　　　　地面层地面　　　　上层地面

楼梯和坡道的表示

在设计图纸上,楼梯的表示从踏步数开始,接着是踏面和踢面的尺寸(如:10步17.5cm/26cm)。楼梯的上下方向用连续线来表示,起始端用圆圈表示,终止端用一个箭头表示。斜坡用两条斜线表示,由坡道起始端到终止端的中点。

在剖面图上,楼梯结构应尽可能简单表示,以便能轻松了解楼梯的几何尺寸:其中包括区分混凝土楼梯和木(或金属)楼梯,楼梯踏步是封闭的还是敞开式的,以及是否有休息平台等。

平面图的尺寸标注

设计制图的尺寸标注用于说明建筑物和室内尺寸的一致性,同初步设计图纸一样,包括外部所有覆盖层和抹灰层的外部标注是第一道尺寸,这就很容易确定楼层的粗略面积和建筑物在规划图或者位置图上的空间位置。

第二道尺寸标注了外部的门和窗户,附加的尺寸线表示窗的内部位置(见图39)。因此,在建筑物的立面图上,所有孔洞的几何关系及对外墙上的空间影响都可以表示出来,内外轴线任意变换或者任意窗槽在图纸上都可表达出来。

接下来的一道尺寸标注了内部空间的长和宽,这对计算内部空间

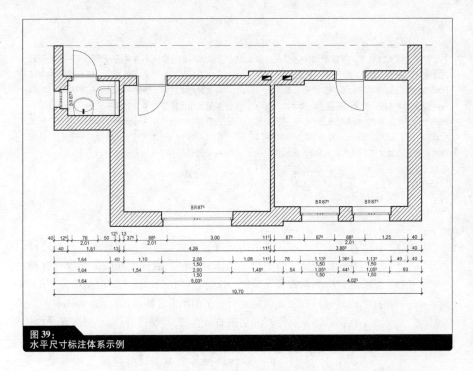

图39:
水平尺寸标注体系示例

面积和房间尺寸非常重要,也有利于以后设备布局。所有的房间尺寸、墙体尺寸及上面的门洞尺寸等都应单独标注出来。在尺寸线上附加数字说明孔洞的高度(门,窗台和窗户),窗台高度标注为窗户。制图者应该注意到尺寸布局的合理顺序,以便于图纸的理解。民用建筑图纸的尺寸的典型布局从外到里依次是:

第一道尺寸:外部总体尺寸线(可利用的位置,附加尺寸线标注凸起和凹进墙体);

第二道尺寸:所有外部孔洞尺寸(门,窗户和突起等);

第三道尺寸:所有墙体孔洞内部尺寸;

第四道尺寸:外墙内部的房间尺寸;

第五道尺寸:房间的内墙,门,凹进部位,拐角和走廊尺寸;

第六道尺寸:内部房间的房间尺寸;

第七道尺寸:其他内容。

对于诸如工业建筑的设计,在平面图上轴线的尺寸标注布置在图的最外侧,并以连续数字从左或从右依次表示轴线编号,轴线编号写在图的顶部或底部。

> **提示：**
> 在砌体结构建筑中，可以考虑采用"八步格尺寸标注"来控制施工图尺寸。砖长包括一个1cm宽的节点在内共24cm。假如没有表示出节点，则用施工尺寸标注（11.5cm；24cm；36.5cm；49cm 等），如果给出节点，则用特殊尺寸标注（12.5cm；25cm；37.5cm；50cm 等）参见本套基础教材中的《砌体结构》，余流译，中国建筑工业出版社 2010 年出版（征订号：18859）。

> **提示：**
> 结构的尺寸线必须考虑仔细，尽可能不要进行二次标注，也不要遗漏标注。假如建筑物内部有大量的小房间。想要标注出所有尺寸非常麻烦。如果单个尺寸不能够满足要求，那么可以直接在内部增加附加尺寸（平面图内）。这比在整个建筑图纸外侧加上一道尺寸，或十分成局部尺寸线更容易处理。

在平面图上不仅要给出长度和宽度，也要给出标高。只有这样，才能在绘图或施工时给出平面图的标高。假如平面图上面没有标高，那么这些标高常常在入口处一次标注。在剖面图上，会给出详细的标高。因此，只有剖面图完成后，平面图上才能给出标高。

剖面尺寸标注

在剖面图上，标高和层高非常重要。同时，标高和层高在平面图上也需要标注。常使用标高符号标注，并作为线性尺寸的补充（见"表现原则，尺寸标注"章节）。标高符号显示了绝对高度（相对零水平面），线性尺寸给出了个别建筑和房屋的高度。例如：屋顶标高表示出承载结构的高度和屋顶结构的总体高度（脊高度，阁楼高度）。在层高表示时，需要区分以下各方面：

— 层高：从这一层的顶部到上一层的顶部；
— 净高：这一层的地板上侧到上一完整层的底部（可利用的位置，抹灰地面层或者吊顶边缘）；
— 建筑高度：该层的结构层上侧距离上一层结构层底侧。

以下是线性尺寸标注的布置（同平面图）：
第一道尺寸：外部总体尺寸线（可利用的位置，附加尺寸线标注凸起和凹进墙体）；
第二道尺寸：所有外部孔洞尺寸（门、窗户和凸出部位等）；
第三道尺寸：所有孔洞内部尺寸（门、窗户和凸出部位等）；
第四道尺寸：净高；
第五道尺寸：其他。

立面图尺寸标注

立面图上采用标高符号标注。而对于更大的建筑，也需要采用线性尺寸标注来作为补充。

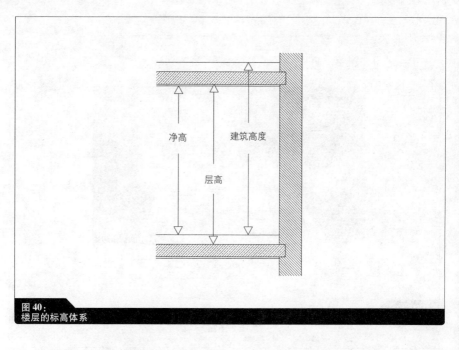

图40：
楼层的标高体系

在剖面图上，标高起点在地面层完成面的水平零线上，在水平零线上还可以添加其他的附加信息。

有时，应当加入水平零线的附加信息，尤其是在表示街道标高和地形坡度时。

文字标注

通常情况下，在设计阶段，只有那些显著的构件才会给出标高，如地表面和屋脊。

在平面图上，需标记出房间名称，名称内容包括房间号或者房间名称（厕所，起居室等）和房间使用面积。用实心黑色三角形表示在建筑进口，便于更容易区别。

上坡和下坡等倾斜面用高度尺寸和角度表示，在剖面上用百分率或者度数并附带方向的箭头补充来表示（例如：房屋沥青层向右45°）。画上一个北向箭头，表示各个房间的光线朝向性，以确定建筑立面。

P50

设计许可

在这个阶段，根据与制图相关的权威规范，对场地位置图和设计图纸补充附加信息。根据规划建筑的性质和规模，需要强制加入

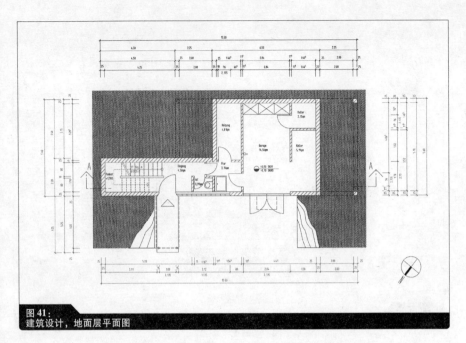

图 41：
建筑设计，地面层平面图

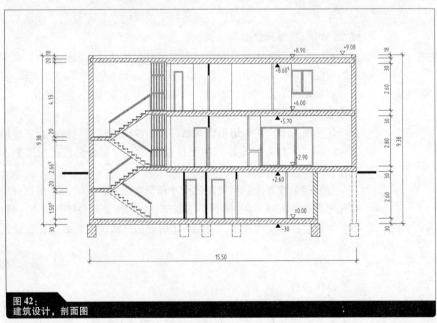

图 42：
建筑设计，剖面图

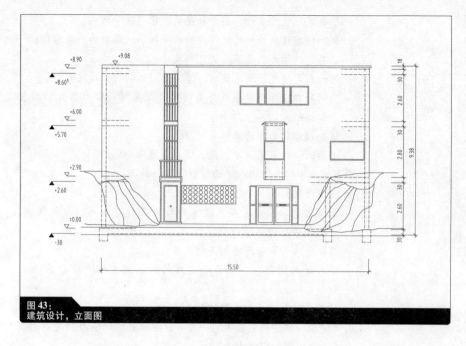

图43：
建筑设计，立面图

各种要求。原则上，设计图纸可以很容易地转换为上交的设计图许可书。

官方的场地位置图

场地位置图通常是由公开指定的测量员或者当地测量部门所绘制得到。在提交设计许可申请时，需要权威人士对所需的要求给出解释。对于一份官方的场地位置图，需要区分文字部分和设计部分。场地位置图的比例通常为1：500，但是对于很大的项目或很小的项目，也可以选用1：1000或者1：250的比例。场地位置图常以黑白色表示，但对于可用范围的边界和面积，也可选用其他的颜色。

场地位置图应该包含以下内容：

— 建筑场地相对于定向点的位置，指北针；

注释：
如果在此阶段进行计算，那么从度数到百分率的转换是简化的：一个斜度X%表示水平位置100cm，标高变化xcm。如果大量的度数被转化为百分率，转化等式是：正切角＝对边/相邻边。例如，度数为10°，tan10°＝0.1584。相应的就是水平位移100cm，标高变化15.84cm，即15.84%。

—已有建筑物的名称，层数和屋面形状（屋脊方向）；
—规划的建筑物外部尺寸，相对水平标高，层数和屋面形状；
—已有建筑物和新建筑物的外部尺寸；
—场地利用情况，如：花园，停车场，运动场，庭院等；
—表明和确认相邻场地到公共场所的距离（通常以单独的图表示相隔距离）；
—限制性建筑面积的标示和划分；
—必要的管线布置（水，电，暖，无线电/电话线）。

以下的文字信息用来对场地位置图进行补充：
—绘图比例；
—街道名称和房屋号码，屋主和场地名称（边界，开放区域，分割部分）的详情；
—面积标注，注册场地相关的边界；
—现有树木的信息，尤其是受自然保护或者树木保护条令的树木；
—建筑物限制的面积和用途信息。

假如建筑场地是规划发展的一部分，它就必须严格遵照规划所要求，这些通常以图片形式表示。

已有建筑改造

假如设计的不是新建筑，而是扩建或改建，在图纸上必须明确区分开拆除部分和新建部分。通常，在项目开始之初，就需要绘制完整的图纸，对改建或拆除措施给出计划。假如拆除量大，则还需要提供单独的拆除阶段图纸，甚至更详细的施工图纸，作为拆除工作的依据。在平面图和剖面图上，用斜度45°的正交虚线表示拆除区域。另外，在已有建筑物中，拆除部分和新建部分在平面图上用颜色表示：

黑色：需要保留的旧的、已有截面；
红色：新增截面；
黄色：在改造过程的拆除部分。

总的来说，理想的设计图纸应当经过设计权威的讨论。

排水图

排水图表示出了排水管道的路线，如厕所和厨房卫生设备的管道和外部屋檐表面的排水管道。在平面图和剖面图上，给出了落水管和设备的连接管道。管道通常由管径确定（例如 $DN\ 100$ 表示：内径100mm），根据管径，可以给出墙上的安装固定件或井道水泵等的尺寸。

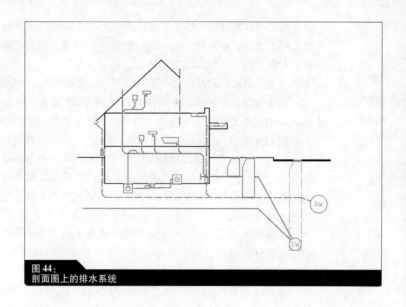

图44：
剖面图上的排水系统

P52
施工图的
目的

施工图

施工图的目的是为精确地建造建筑物提供信息。因此，施工图必须包含所有与建筑物完工相关的分类信息，包括其他专业人士完成的建筑设计、暖通、卫生管道、承重结构、防火等。

施工图根据绘图比例和详细程度可分两个部分：施工图部分，比例为1:50；详图部分，比例从1:20~1:1，以下是两部分施工图的基本差别。在此方面，没必要用上两部分间所有的差别。

— 比例为1:50的施工图包括平面图、立面图和剖面图，这些都可以表达建筑物整体或其中一部分；

— 正剖面图的比例为1:50~1:10，通常情况下，正剖面图通过断面、内外视角等给出更为详细的内容，以此给定建筑物其他部分在结构和几何上的关系；

— 安装图纸的比例为1:50~1:20，个别设备需要画专门的安装图。这些包括抹灰、贴面、地面装修、干法施工、吊顶等；

— 详图的比例从1:20~1:1，详细给出了特殊结构、构件连接等的情况；

— 设备现场安装图。

施工图的
表现方法

制图者必须把设计转化为合理、完整的施工图、详图和施工方案图，并且，所选用的方式应使此工程的施工人员能轻易地读懂和操作。

施工图必须非常精确，以至于在说明和尺寸中不允许出现任何意外情况。但也没有必要定义出每个螺钉的尺寸，因为可以假定施工者有必要的专业知识。规划部门拟定的需求也都要在施工图上反映出来。施工阶段如果出现变动和模糊不清，施工图必须紧接着给出相应调整。

相交结构构件的表示方法

通常情况下，相交的结构单元以比例 1∶50 来表示，以便能够直接表示出墙体结构的结构信息（例如，砌体墙两侧的水泥砂浆、混凝土顶棚的抹灰层）。墙上和顶棚的洞口和切口也应表示出来。这些洞口和切口常用来安装设备（例如，壁炉，暖通管道，卫生设备，通风和电器设备），并且得到设计者同意。切口用对角线来表示；在结构构件被完全剖切后，洞口是封闭的。为了显示其重要性，也用实心的黑色三角来表示。

施工图的轴线

承重结构体系重复的建筑，例如工业用房或者办公大楼，承重的纵横框架轴线以数字或字母表示。还可以对个别区域进行直接分类，以便于结构工程师能够利用定义的轴线联系到建筑物的系统性和功能。轴线以点划线表示，既能整个画出来，也可以只画在建筑的外部。

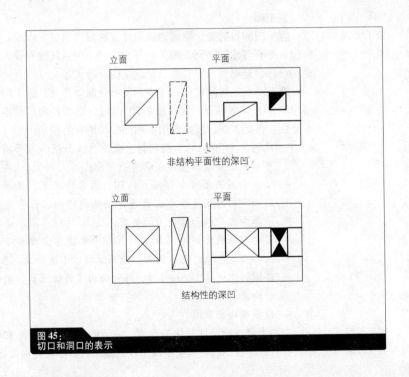

图 45：
切口和洞口的表示

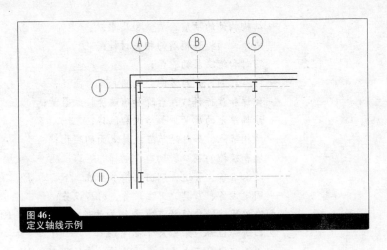

图46:
定义轴线示例

施工图的尺寸标注

在这个阶段上,所有的尺寸标注必须正确地表示出来。这不仅包括设计图纸上已标注的所有尺寸;而且也包括相关结构构件的标高、宽度和深度,缺一不可。每个构件必须附属在建筑物的一部分,以便进行现场施工测量。在平面图和剖面图上,设计图纸中的常规尺寸首先标注在外侧,而详细的尺寸则标在内侧。

房间的文字标注

在施工图中,房屋的文字标注比设计图纸上要包含更多的信息。除了房间号和房间编号外,还包括:

— 房间面积(A),单位:平方米;
— 周围墙体(S)信息,例如踢脚板的尺寸;
— 净高(CH),为了满足接下来的墙体施工需要,例如,粉刷尺寸。

同时,也可以尽可能标记出特殊的楼层、墙体和顶棚安装。如果房间内部大小不能满足标注的需要,则可以在设计总说明中交待(见"图纸的表现方法"一章),而相应的房间可以用缩写字母来表示(例如,用W1,W2标识墙体上的设备等)。

施工图

平面图的内容

平面图应该有以下信息:

注释:
标注时,首先应把自己想像成一个现场施工者,必须按图纸进行施工。例如:安装一扇门时,如果门的外框与外缘相接,那么门的中轴线标注就没有什么用。同时,也需要尽量避免双重标注,因为这增加了相互间转换的工作量;另外,双重标注也容易被忽略,而产生图纸的前后不符。

— 结构构件的性质，质量和尺寸；
— 墙体，顶棚和楼层的相关材料；
— 密封和绝缘层的表示；
— 门窗孔洞，开启方向，孔洞和窗台高度；
— 楼梯和步行斜坡，台阶和跨越阶详细数量；
— 结构单元的质量，如防火和隔声；
— 结构节点，如外延节点或外表面的变化；
— 墙和顶棚的洞口、切口、电梯井等；
— 技术设备，管道，烟囱，排水系统，下排水等；
— 固定设备、家具、卫生设备，厨房设备；
— 准确地建造建筑物，需要结构构件的所有标注（任何凸起或凹进部位都应给出尺寸标注）；
— 用于确定房间大小和数量计算所需的所有尺寸；
— 房间标注（如上所需）；
— 基于零水平面的标高，由此能清晰表示楼层层高；
— 详细参考资料。

立面图和剖面图的内容	立面图和剖面图包含以下信息： — 层高，净高，结构标高； — 板层，地板，基础和屋顶等的标高； — 地板和屋面结构； — 表示已有或已规划的地面情况； — 门、窗图形和分栏； — 排水沟，落水管，烟囱和屋面结构； — 用虚线表示的被遮挡的中间楼层，承力墙和基础； — 基坑开挖过程； — 顶棚，屋面，楼层结构详图； — 如果立面图上的玻璃规格不同，需要标明。

立剖面图

为了详细地表现出整个立面情况而不是个别的细部，立剖面图是值得绘制的。如此一来，剖面图中能完整地给出标高设定、内视图、外视图。同时，通过内部和立面在标高和所有连接点之间的关系，可以对平面图进行详细补充。

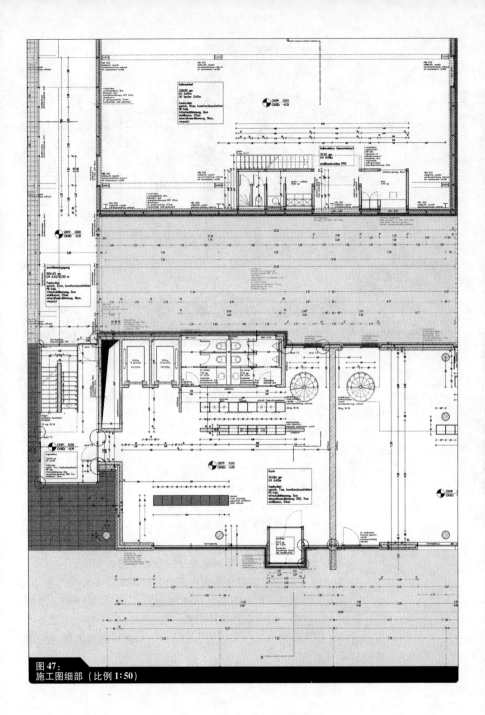

图 47:
施工图细部(比例 1:50)

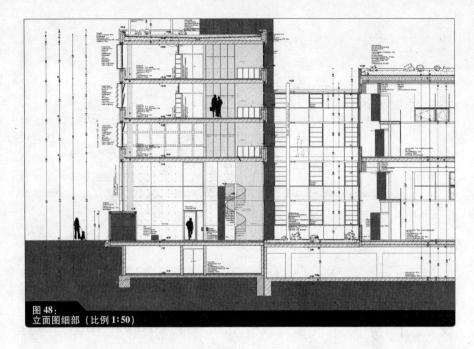

图48：
立面图细部（比例1:50）

设备安装图

设备安装图是展示特殊结构构件安装的施工图纸。它包含以下结构构件的安装信息：

— 预制钢筋混凝土；
— 钢结构；
— 木梁或屋架；
— 打磨部分（伸缩缝和导管）；
— 石料铺设（网格划分，交叉点和伸缩缝）；
— 表面贴砖（网格划分，配件及伸缩缝）；
— 顶棚吊顶（网格划分，配件，隔声效果区等）；
— 叠合楼板或空心楼板（网格划分，地板下内部装修）；
— 楼板装修层（网格划分或轴线布局，楼面装修层的变化等）。

安装图纸常常基于现有的施工图进行绘制，并辅以其他的适当信息，例如：网格线，颜色和填充阴影线。安装图包括了特殊的服务区域，且通常是在管理人员介入之前绘制完成，如此可以将其作为管理条例的附属部分。

图49：
立剖面图示例（比例1:20）

详图

详图包含了所有种类的连接、结构体系和转换部分的内容。如同设备安装的绘制方法，在详图中，有很多重要的点是多种控制系统相交或汇聚的地方。由于一个工程的详图设计依赖于自身特点、对详图设计的深度要求、设计要求以及施工单位的问题或不确定地方等多种因素，因此不可能给出详图设计的普遍规定。需要详图的区域包括：

— 立面墙：窗户节点，构件和地面和立墙的过渡部分，外墙和屋顶的连接，拐角位置，外门，阳台，窗台，遮阳罩；

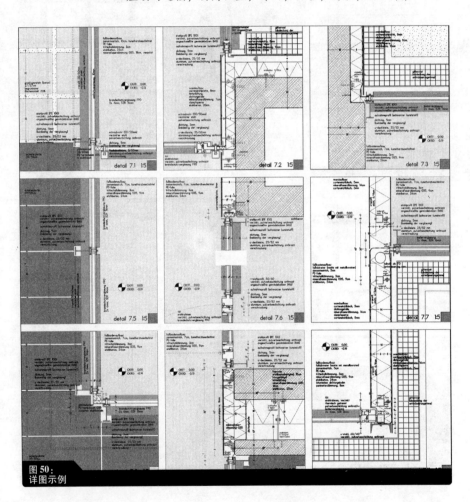

图50：
详图示例

— 地基：地基，排水，密封，绝缘地面；
— 屋面：阁楼，屋檐，屋脊，边缘，山墙，屋顶开孔，如烟囱，通风，天窗和屋顶窗；
— 楼梯：结构剖面，上下连接，休息平台，楼梯栏杆，扶手；
— 楼面和顶棚装置：所有顶棚装置轮廓，各楼层的转换，建筑物立剖面图上的连接，设备和管道；
— 门：系统门，框架结构体系，钢框门，电梯门，转轴门；
— 干法施工：墙体外立面的连接，结构完成面，楼面及顶棚，顶棚吊顶；
— 厕所，厨房，家具：结构详图，厕所隔墙等。

现场施工图

现场施工组织图用于协调现场和施工。对于小工程，通常没有必要单独画现场施工图。但是如果现场空间狭小，就有必要用现场施工图来避免相互影响而造成场地利用率低下，因此，现场施工图应记录以下部分内容：

— 存放区和工作区/工地路；
— 现场管理处；
— 住宿和卫生设施；
— 工作周围地区建筑；
— 基坑开挖；
— 起重设备（如起重机）的工作半径和业务范围；
— 工地的围墙，大门，标志等；
— 特殊工种区域（如与混凝土施工相关的钢筋弯曲和切割区域）；
— 合适的堆土放置区域；
— 供电和供水设施，处理设施，垃圾管理等。

施工安装图

不同职能的施工单位以施工图为基础，把其融入自己的施工安装图中（有时候也叫施工工艺图）。在动工之前，这些图纸必须交给设计者审核并签字。典型的专业施工单位的施工安装图包括以下部分：

— 金属或者钢结构部分（窗，钢结构建筑，栏杆等）；
— 木材或者木工品（木结构，屋顶桁架，窗等）；

— 通风施工；

— 电梯施工。

P63

专业图纸

承重结构

面向结构工程师的施工图

结构工程师画出自己的施工图纸，重点关注那些与静力学相关的内容。需要绘制什么样的图纸，主要依赖于所选用的建筑材料。如果选用钢筋混凝土，那么应当包括模板图和钢筋图；如果选用木材或者钢材，那么应当包括檩条图、木料图和钢结构图。

位置图

位置图标出了用于说明静力计算的一些特殊点。根据设计图纸对这些点进行编号，静力计算也可以找到这些编号。

模板图和钢筋图

在钢筋混凝土建筑中，应绘制模板图和钢筋图。这里，模板图表现出装入套筒的结构单元（例如，混凝土顶棚和墙体）。假如以后视觉效果上非常重要，那么模板图所表示的就是构件的结构完成部分

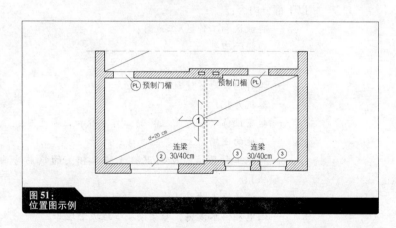

图 51：
位置图示例

提示：
结构工程图通常不能表现出下部楼面的俯视图，但能够表现出相应的顶棚部分。在结构工程平面图上，第 2 层平面图以及不可见的结构相交构件，表现出与建筑师的设计完全不同的视点。为了说明，你可以想像成一个镜面地板，上面显示了顶棚的所有外轮廓线。

例如：
一个混凝土梁静力计算分析时，分配的号码为 21，这根梁在位置图上编号也就是 21，如此在建筑静力计算中可以清楚表达该梁。如果采用这样的表示方法，计算中所有相同的梁可以被编上相同的号码。

（例如，钢筋混凝土的顶棚和墙体）。如果随后的外表面有很重要的视觉效果，则模板图就显得更为重要了（例如，清水面混凝土墙）。

顶棚模板图应给出以下内容：
— 轴线，重量和高度；
— 支撑和承重建筑单元；
— 承重墙上的孔洞；
— 类型和强度等级；
— 跨度方向。

钢筋图上包含了钢筋混凝土构件中钢筋网和钢筋等内容。一片钢筋网通常以斜线填充的矩形面积表示。钢筋图上附加信息如下：
— 钢筋类型；
— 钢筋的数目，直径，形状，长度以及焊接；
— 混凝土强度等级，保护层；
— 洞口和特殊结构；
— 图纸上附带建筑上的钢筋材料表。

木结构施工图

对于木结构，应绘制木结构施工图，给出每根木料的精确位置和尺寸（横梁，支撑和檩条等）。对轴线给以尺寸标注，节点单独给出详图。例如，如果建造了一个斜坡屋面，那么在椽的详图上就必须给出檩条和椽子的位置和尺寸。

建筑设备

建筑或当地设备安装图也应该绘制专图，提供安装方法。每种设备都经过单独设计。需要绘制设备安装图的特殊设备包括：
— 加热装置；
— 供水和污水处理装置；
— 通风装置；
— 电气装置；
— 消防技术及警报器；
— 数据技术设备；
— 电梯设备。

说明：
本套基础教材中的《屋顶施工》给出了详细的建筑木材和椽子的图纸，中国建筑工业出版社 2010 年出版。

如同设备所需的实际空间,如公用服务空间、锅炉房等,管道走向、孔洞和缆线长度是十分重要的特点。当结构完成部分中介入了准确的细节部分,那么建筑图纸中需要包括那些表示了切口和孔洞的图纸。

图纸的表现方法

如果要打印纸质的图,则必须设定图纸型号,以符合现代复制的要求,并且还需给出图签。

绘图面积

图纸的组成

一旦建筑或者剖面上需要尺寸标注,那么图纸的比例就固定了,则所需的图纸面积就可以计算出来。由比例换算得到图形面积,加上两侧标注尺寸所需的空间,就可以得到合适的绘图面积。图签必须在绘图区域旁(见下图),呈矩形,且允许与图纸边缘相交。

选择图纸型号

对于不同目的的图纸(竞赛,学生作业),都可能有相应尺寸的图纸型号。例如,一个窄而长的建筑可以绘制在具有相同比例的图纸上,这样可以有效加深建筑在形状上的效果。

对于建筑图,选择常规的图纸型号是有意义的(例如,DIN 中的 A 型系列),这便于复制。而施工图需要用较大的图纸,但对于详图一般用小的图纸,如 DIN-A 图纸,正好便于在大多数复印机上进行复印。(见"表现原则"一章)

不同的比例

在同一张图纸上,可以有不同比例图,例如,沿着一个立剖面图,以合适的比例来表示个别的锚固点是有意义的。需要注意的是:给出明确的文字标注,以精确识别特别的图。

图签

每张设计图都应有图签,可以清楚说明该图是哪一工程,该张图纸上是何详图内容。图签通常在图的底部右下角,对于效果图和施工图,图签的内容安排有所区别。

效果图的图签

如果图纸是为了表现出设计效果,则应该有一个相应的图签。设计总图通常有图签,内容包括:项目名称,详细比例,图形名称(如地面层图)和作者。在图纸标题上也应有设计图表,截面说明,水平标高和指北针。

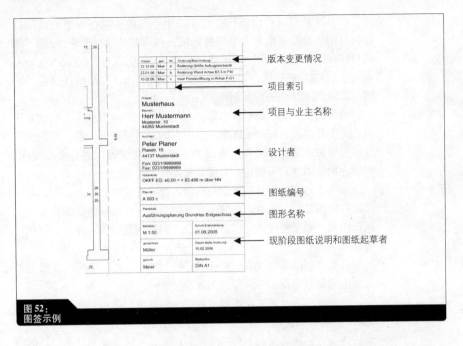

图 52：
图签示例

对于大学生的设计，个人情况应包含学生的入学编号，名字，院系和老师。对于竞赛的图纸来说，不需要名字或者院系，取而代之的是竞赛提供的编号，填写在图纸上一角，提交后密封起来；在竞赛评审以后打开，方可知道设计的作者。

施工图的图签

对于施工图纸，图签首先是客户名字、设计者、图纸起草者和比例，并给出现阶段图纸说明。对于新型或者专门的设计材料，如果需要定期使用，则必须记录当前图纸的准确信息。一种较好的记录方法是：给出第一版和当前版的日期。这也有利于理解图纸版本的变更过程，图纸的总结，附加信息等。建设许可图有图纸编号，定义得非常清楚。为了简化，当前图纸是在图纸编号里的一部分，图纸 A34d 表示：施工图第 34 号，第四版（d）。当然，图纸编号可以随意分配；但如果在工程开始前就建立编号，且与所有相关的内容协调一致，那么对于图纸的管理来说就非常容易了。

P68

图纸布局

现在，CAD 程序的使用已非常普遍，很少有设计者采取手工画图了。二者相比，CAD 程序绘图更容易复制和图纸布局。

127

打印　　　　CAD 文件可以在绘图仪上打印，小号图纸可以在标准的商用打印机打印。绘图仪作为大型的打印机，可以打印 A1 或 A0 号图，图纸打印在滚筒纸上面。A1 滚筒纸的宽度是 61.5cm，A0 是 91.5cm。在大多数 CAD 程序里，可以单独设定图纸尺寸，这使得任意型号的图纸都能与滚筒纸的宽度相匹配。

复制　　　　假如需要复制一份已经打印出来的图纸，则可以利用大图的复印机。大图复印机装配的纸张常与绘图仪的滚筒纸宽度一致。彩色大图复印机通常价格比较昂贵，所以，一张图纸应该可以进行多次复印。对于详图的图纸，可以用标准商业复印机来复印。

蓝图　　　　如手工绘图是在透明纸上完成的，则有可能以此得到蓝图，而不需进行大图复制。此时，将蓝图放在原始的、绘制好的透明图上，利用蓝图机上的紫外线来完成。蓝图有不同的颜色类型，黑色、红色或者蓝图。

　　　　目前，蓝图非常少见，CAD 制图已取而代之。但是，蓝图为手工制图的复制提供了一种经济的方法。

折叠　　　　许多项目中，图纸在流通过程中不再是他们最初的图纸大小，而是 A4 的型号，这意味着比较大型号图纸的图要折叠到 A4 尺码。图纸折叠后，必须确保右下角的图标露出来，可以被看到。在归档的时候图纸也要尽可能展开，所以装订图纸的孔洞只能在左下角。

> 注释：
> 　　复印店里有机械折叠机，可以很方便地批量折叠图纸。如果手工折叠图纸，最好有一张 A4 的纸板，首先，将左手边图纸折叠成 A4 的宽度，利用纸板折叠摁平，然后右手边折叠大约 19cm 的宽度，直到剩余能被折叠完后再折叠底边部分到 A4 的高度（对于大号图纸，需要多次），在内部拐角，正如图表里显示的一样，应当向内折叠，因此不利于进行冲孔装订。如果你使用了图签或 CAD 打印中的矩形图签，那么就不可能在图框内画线了，因此也没有必要进行测量，图纸能够直接进行沿线折叠。

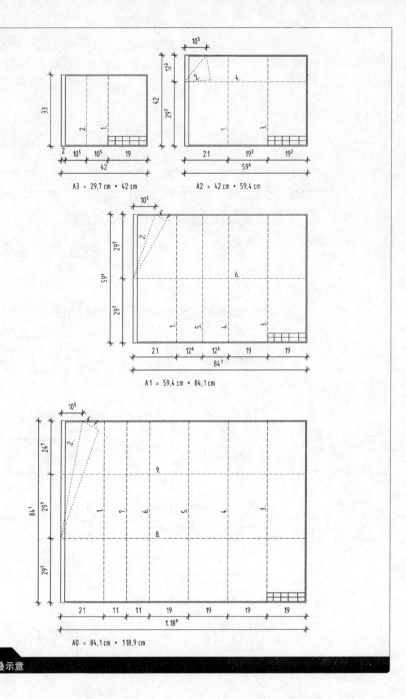

图 53:
图纸折叠示意

附录

符号

卫生设备的符号 表4

项目	符号
洗手盆	
水池	
水箱式便池	
无水箱便池	
机械式喷洗座便器	
小便器	
沐浴盆	
浴缸	
洗衣机	WM
滚筒烘衣机	
浴缸角落	

厨房设备的符号　　表5

项目	符号
下柜	
上柜	
上下柜	
沥水板式洗涤池	
煤气炉	
电炉	
一体式煤气炉烤箱	
嵌入式烤箱	
厨房操作面	
电冰箱	
冷冻箱	
洗碗机	
微波炉	
抽油烟机	
椅子	
餐桌	

家具设备的符号 表6

项目	符号
厨柜	
扶手椅	
沙发	
桌子	
椅子	
三角钢琴	
直立式钢琴	
书桌	
衣柜	
单人床	
带有床柜的单人床	
双人床	
带有床柜的双人床	

标准

工程制图按照国际标准,因此,很多国家的制图格式是一样的,以下给出了合法的标准化标准,这些标准由德国国家标准化机构(例如:ISO 128 在德国称为 DIN ISO 128)制定。

ISO 相关标准　　　　　　　　　　　表7

ISO 标准	说明
ISO128	工程制图。表现总则
ISO216	书写纸和图纸分类。裁剪尺寸。A 与 B 系列
ISO2594	建筑制图 – 投影方法
ISO03766	施工图。混凝土加固的简化表现方法
ISO4067	建筑与市政工程制图 – 安装
ISO4069	建筑与市政工程制图 – 剖面视图表现 – 总则
ISO4157	施工图 – 命名系统
ISO5455	工程制图。比例
ISO5456	工程制图。投影方法
ISO6284	施工图。界线变化的表示
ISO7518	施工图。拆除与重建的简化表现
ISO7519	施工图。大体布局和安装图表现总则
ISO8048	工程制图。施工图。视图、剖面和剪切的表现
ISO8560	施工图。模数尺寸、线条和栅格的表现
ISO9431	施工图。绘图和文字、标题等在图纸上的空间位置
ISO10209	工程产品存档。词汇。工程制图相关术语
ISO11091	施工图。景观绘图实践

尊敬的读者：

感谢您选购我社图书！建工版图书按图书销售分类在卖场上架，共设22个一级分类及43个二级分类，根据图书销售分类选购建筑类图书会节省您的大量时间。现将建工版图书销售分类及与我社联系方式介绍给您，欢迎随时与我们联系。

★建工版图书销售分类表（见下表）。

★欢迎登陆中国建筑工业出版社网站www.cabp.com.cn，本网站为您提供建工版图书信息查询、网上留言、购书服务，并邀请您加入网上读者俱乐部。

★中国建筑工业出版社总编室
 电　话：010—58934845
 传　真：010—68321361

★中国建筑工业出版社发行部
 电　话：010—58933865
 传　真：010—68325420
 E-mail：hbw@cabp.com.cn

建工版图书销售分类表

一级分类名称（代码）	二级分类名称（代码）	一级分类名称（代码）	二级分类名称（代码）
建筑学（A）	建筑历史与理论（A10）	园林景观（G）	园林史与园林景观理论（G10）
	建筑设计（A20）		园林景观规划与设计（G20）
	建筑技术（A30）		环境艺术设计（G30）
	建筑表现·建筑制图（A40）		园林景观施工（G40）
	建筑艺术（A50）		园林植物与应用（G50）
建筑设备·建筑材料（F）	暖通空调（F10）	城乡建设·市政工程·环境工程（B）	城镇与乡（村）建设（B10）
	建筑给水排水（F20）		道路桥梁工程（B20）
	建筑电气与建筑智能化技术（F30）		市政给水排水工程（B30）
	建筑节能·建筑防火（F40）		市政供热、供燃气工程（B40）
	建筑材料（F50）		环境工程（B50）
城市规划·城市设计（P）	城市史与城市规划理论（P10）	建筑结构与岩土工程（S）	建筑结构（S10）
	城市规划与城市设计（P20）		岩土工程（S20）
室内设计·装饰装修（D）	室内设计与表现（D10）	建筑施工·设备安装技术（C）	施工技术（C10）
	家具与装饰（D20）		设备安装技术（C20）
	装修材料与施工（D30）		工程质量与安全（C30）
建筑工程经济与管理（M）	施工管理（M10）	房地产开发管理（E）	房地产开发与经营（E10）
	工程管理（M20）		物业管理（E20）
	工程监理（M30）	辞典·连续出版物（Z）	辞典（Z10）
	工程经济与造价（M40）		连续出版物（Z20）
艺术·设计（K）	艺术（K10）	旅游·其他（Q）	旅游（Q10）
	工业设计（K20）		其他（Q20）
	平面设计（K30）	土木建筑计算机应用系列（J）	
执业资格考试用书（R）		法律法规与标准规范单行本（T）	
高校教材（V）		法律法规与标准规范汇编/大全（U）	
高职高专教材（X）		培训教材（Y）	
中职中专教材（W）		电子出版物（H）	

注：建工版图书销售分类已标注于图书封底。